T0302118

FORMAL LANGUAGES IN LOGIC

Formal languages are widely regarded as being, above all, *mathematical objects*, and as producing a greater level of precision and technical complexity in logical investigations because of this. Yet defining formal languages exclusively in this way offers only a partial and limited explanation of the impact which their use (and the uses of formalisms more generally elsewhere) actually has. In this book, Catarina Dutilh Novaes adopts a much wider conception of formal languages so as to investigate more broadly what exactly is going on when theorists put these tools to use. She looks at the history and philosophy of formal languages, and focuses on the *cognitive* impact of formal languages on human reasoning, drawing on their historical development, psychology, cognitive science, and philosophy. Her wide-ranging study will be valuable for both students and researchers in philosophy, logic, psychology, and cognitive and computer science.

CATARINA DUTILH NOVAES is Assistant Professor at the Faculty of Philosophy of the University of Groningen, The Netherlands. She is the author of *Formalizing Medieval Logical Theories* (2007).

FORMAL LANGUAGES
IN LOGIC

A Philosophical and Cognitive Analysis

CATARINA DUTILH NOVAES

CAMBRIDGE
UNIVERSITY PRESS

CAMBRIDGE
UNIVERSITY PRESS

University Printing House, Cambridge CB2 8BS, United Kingdom

One Liberty Plaza, 20th Floor, New York, NY 10006, USA

477 Williamstown Road, Port Melbourne, VIC 3207, Australia

314-321, 3rd Floor, Plot 3, Splendor Forum, Jasola District Centre, New Delhi - 110025, India

79 Anson Road, #06-04/06, Singapore 079906

Cambridge University Press is part of the University of Cambridge.

It furthers the University's mission by disseminating knowledge in the pursuit of
education, learning and research at the highest international levels of excellence.

www.cambridge.org
Information on this title: www.cambridge.org/9781107020917

First published 2012
First paperback edition 2014

A catalogue record for this publication is available from the British Library

Library of Congress Cataloging in Publication data
Novaes, Catarina Dutilh.
Formal languages in logic : a philosophical and cognitive analysis / Catarina Dutilh Novaes.
p. cm.
ISBN 978-1-107-02091-7 (hardback)
1. Formal languages. 2. Reasoning. 3. Cognition. I. Title.
QA267.3.N68 2012
511.3–dc23
2012015483

ISBN 978-1-107-02091-7 Hardback
ISBN 978-1-107-46031-7 Paperback

Contents

Acknowledgements   *page* vii

Introduction   I

PART I   9

1 Two notions of formality   II
   1.1 The formal as de-semantification   12
   1.2 The formal as computable   16
   1.3 Conclusion   28

2 On the very idea of a formal language   29
   2.1 'Language' and languages   29
   2.2 What are formal languages then?   52
   2.3 Conclusion   65

3 The history, purposes, and limitations of formal languages   66
   3.1 The historical development of a technology   66
   3.2 What are formal languages good for?   89
   3.3 Pitfalls and limitations   97
   3.4 Conclusion   108

PART II   III

4 How we *do* reason – and the need for counterbalance in science   113
   4.1 'Disproving Piaget'   114
   4.2 'A fundamental computational bias'   129
   4.3 A pluralistic conception of human rationality   147
   4.4 Conclusion   159

5   Formal languages and extended cognition   161
    5.1   Reading in general   162
    5.2   Formal languages and extended cognition   178
    5.3   Conclusion   196

6   De-semantification   198
    6.1   De-semantification and re-semantification   199
    6.2   Two case studies   211
    6.3   Conclusion   219

7   The debiasing effect of formalization   221
    7.1   Formalizing target phenomena   223
    7.2   Debiasing and cognition   236
    7.3   Conclusion   248

    Conclusion   249
    Different methodologies in philosophical investigation   250
    What each of them has to offer   254
    How they can be combined   255
    The different methodologies in the present work   256

References   258
Index   273

Acknowledgements

This book was for the most part written during my period as a postdoctoral researcher at the Institute for Logic, Language and Computation (ILLC) in Amsterdam (2007–11), on a VENI grant from the Netherlands Foundation for Scientific Research (NWO). So, first and foremost, I would like to thank NWO for their generous support. Further, I want to thank my former colleagues at the ILLC, who provided such a fruitful environment for the development of this project, in particular: Benedikt Löwe, Chantal Bax, Dora Achourioti, Edgar Andrade-Lotero, Erik Rietvelt, Floris Roelofsen, Frank Veltman, Johan van Benthem, Katrin Schulz, Maria Aloni, Michiel van Lambalgen, Paul Dekker, Peter van Ormondt, Remko Scha, Sara Uckelman. Martin Stokhof deserves my deepest and eternal gratitude for his invaluable guidance and support as my 'boss and mentor' during this period. And thanks also to my new colleagues in Groningen, who gave me such a warm welcome in July 2011.

I would like to thank Hilary Gaskin and Anna Lowe at Cambridge University Press for their help at different editorial stages of the project, and Charles Stewart for his vital assistance at the final stages of preparation of the manuscript. Thanks also to the ETH-Zurich Library for granting me permission to use a page of the manuscript of Gentzen's doctoral dissertation for the cover of the book. The page in question has a normalization proof that never made it to the printed version of the dissertation, thus remaining unknown until recently. The theorem was then independently proved thirty years later by Dag Prawitz, and constitutes one of the most fundamental results in proof theory. To me, this page is a beautiful illustration of the power of a suitable notation (in this case, natural deduction) to give rise to groundbreaking results.

In November 2011, the Department of Philosophy at Universidade Federal do Rio Grande do Sul in Porto Alegre (Brazil) hosted a prepublication, author-meets-critics event on the manuscript of this book. It was a real privilege to have my work carefully read by such sharp commentators

viii   *Acknowledgements*

while there was still time to make changes to the end result (the remaining shortcomings are of course my entire responsibility). I thank Alexandre N. Machado, Luis Carlos Pereira, Eros Carvalho, and Paulo Faria for their insightful comments, and Alfredo Storck and Lia Levy for organizing the event.

I am much indebted to colleagues around the world who in some way or another contributed to the writing process along the way, be it in the form of discussions or by commenting on previous drafts: Albrecht Heeffer, Andrew Fugard, Arianna Betti, Danielle Macbeth, David Over, Eric Schliesser, Frederico Dutilh Novaes, Graham Priest, Greg Restall, Hannes Leitgeb, Helen de Cruz, Jeff Ketland, Jeroen Bruggeman, Julien Murzi, Juliet Floyd, Juliette Kennedy, Keith Stenning (who provided detailed comments on the whole manuscript!), Marije Martijn, Markus Schlosser, Ole Hjortland, Rafael Nuñez, Richard Menary, Sebastian Sequoiah-Grayson, Shira Elqayam, Stephen Menn, Stephen Read, Stewart Shapiro, Thomas Müller, Wilfrid Hodges, Wilfried Sieg.

But, crucially, I want to thank Arianna, Dora, Greg, Jeroen, Seb, Stephen, and especially Edgar, Marije, and Ole also for their friendship and support over the years.

Many people ask me how I manage to combine an academic career with family life and children. Well, the short answer is, by relying on an extensive net of loving, dedicated allo-parents. I have been very fortunate to be able to count on the support of fantastic, caring people who have been involved in the rearing of my children over the years, at different times: their wonderful teachers at school, and at home, Camila, Irene, Rafa, Alice, Cris, Ana Célia. Within the family, my mother Maria, my brother Frederico, and Gê have been a constant presence, albeit from far away; my parents-in-law Arent and Denise are constantly involved in my children's lives – Denise must be the only mother-in-law in the world to have accompanied a daughter-in-law to an academic conference in order to care for the children. Without these fantastic allo-parents, this book would never have come into existence.

Sophie and Marie deserve thanks too, for filling my life with joy and trouble, laughter and worries, but most importantly with love. Children and books belong in the same category in that they project one's being out into the future (in the wise words of my friend Greg). But parenting is much harder than writing books, and I can only hope they will be proud of me both for my book-writing and for my parenting skills when they grow up.

But my final thanks should go to Reinout, my biggest fan and supporter. You make me feel like the most amazing person in the world; I could never thank you enough for this and so many other wonderful things.

Introduction

The starting point for this investigation was the almost trivial observation that the output of research in logic is fundamentally different depending on whether it is conducted with or without formal languages. Of course, some might say that research in logic conducted without formal languages is not even 'logic' properly speaking. But the study of the history of logic suggests that such a view is at best limited and at worst misguided: many logical traditions which do (did) not make extensive use of formal languages often display a level of conceptual sophistication that leaves nothing to be desired relative to modern developments. And yet, they *are* very different from current research in logic, and this fact itself calls for an explanation. After all, what is so special about formal languages and formalisms more generally? Whence the magic?

Typical views on formal languages are based on the premise that they are, above all, *mathematical objects*: a greater level of precision and technical complexity in logical investigations arises because they are precisely defined mathematical objects. Yet it would seem that viewing formal languages exclusively from this point of view offers a very partial and limited explanation of the impact that their use (and uses of formalisms more generally elsewhere) actually has. In the present inquiry, the idea is to adopt a much wider conception of formal languages so as to investigate more broadly what exactly is going on when a reasoner puts these tools to use. Furthermore, most of the arguments presented here generalize to uses of formalisms in other disciplines, including the empirical and social sciences.

Originally, my idea for this project was to look more carefully into the role played by formal languages in the social interactions of logicians, that is, the public sphere of logic as a discipline practised by a community of researchers. However, and as is often the case with research, it turned out that a very different approach, one that I stumbled upon almost by chance, offered a more promising perspective on what is so special about formal languages and their uses in logic: formal languages viewed as *cognitive artefacts* enhancing and modifying an agent's reasoning processes.[1] So the key question became: what happens on a *cognitive* level when an agent reasons with formal languages and formalisms in general, as opposed to reasoning without such devices at her disposal?

To address this question, I rely extensively on empirical data on human reasoning as amassed by research within the empirical sciences of the mind. To be sure, there is still a fair amount of disagreement and many open questions in this body of research; it seems we have not yet fully 'cracked the code' of how human beings reason. Nevertheless, many of the results so far obtained do shed light (as I attempt to show throughout the book) on what exactly happens, on the cognitive level, when a reasoning agent operates aided by these special devices, namely formal languages and formalisms.

However, empirical data from psychology and cognitive science are not sufficient for the formulation of an encompassing philosophical understanding of the role of formalisms in reasoning. As it turns out, another essential element in the puzzle is the *historical development* of this cognitive technology. My earlier work as a historian had taught me that many of the assumptions that we take for granted in our conceptualizations of a given phenomenon are in fact substantive commitments, corresponding to significant theoretical steps made along the way. Now, given that the goal is to question what is generally (and uncritically) assumed to be the case about formal languages so as to attain a deeper philosophical understanding of the phenomenon, historical analysis does offer a privileged perspective. It allows us to re-evaluate each of the steps made in the developments towards a given status quo – in this case, the current situation of ubiquity of formal languages and formalisms for research in logic and elsewhere.

To be sure, one of the conclusions to be reached here is that there is a sense in which they *are* indispensable (or in any case irreplaceable), but this will be argued for on the basis of empirical data rather than uncritically accepted. Indeed, one of the main theses of the book is that the rationales that are usually attributed to formal languages – increased precision and clarity, counterbalancing the imperfections of ordinary languages – fail to capture the real impact of using formal languages in practice. I will argue that, rather than expressive devices, formalisms are, above all, calculative, computing devices.

---

[1] By this I do not want to suggest that the public, social perspective becomes unimportant; in fact, my own account of the emergence of logic and deductive reasoning is inherently tied to public situations of dialogical interaction (see section 4.3). Rather, the change of focus is perhaps best viewed as simply postponing the project of adopting the social perspective to a later stage.

Hence, two key components of this study are, first, the analysis of the history of formal languages, and, second, the empirically informed investigation of the cognitive impact of operating with formal languages and formalisms when reasoning. These two elements will be essential in the search for an answer (even if not *the* answer) to the question of why we do or should use formal languages for research in logic, and more generally why (logical and mathematical) formalisms are such powerful epistemic tools.

The historical and the cognitive perspectives also represent two somewhat opposed yet complementary poles in the investigation: the historical perspective emphasizes what is *contingent* in the developments in question, while the cognitive perspective brings in the biological, *necessary* constraints upon the cognitive make-up of human beings. In effect, the historical development of formal languages can be viewed as a process of *cultural evolution* through which humans developed tools that would allow them to perform certain tasks and solve certain problems more efficiently, against the background of the possibilities afforded by human cognition as biologically determined.[2] Naturally, the search process can only take place within the realm of constraints inherent in the human cognitive apparatus, but, just as in processes of biological evolution, a substantial amount of chance and contingency influences these developments. Hence, it would be a mistake to view the developments towards the current situation of ubiquity of formal languages in logic as inexorable; many contingent, cultural, and historical factors played a significant role, as we will see in Chapter 3.[3] At the same time, this does not entail a form of cultural/scientific relativism; indeed, the view to be defended throughout the book is that the development of formal languages and complex systems of notation constitutes *real progress*.[4] A formalism is a powerful technology that allows humans to reason in ways that would be otherwise virtually beyond their reach;[5] explaining why this is so on a cognitive level is the main goal of the book.

---

[2] In other words, the point is a search for tools leading to an optimalization of human cognition given its inherent constraints. Other examples of such processes to be discussed in the book are the evolution of (spoken) languages towards increasing compositionality (Kirby 2012) and the evolution of writing systems towards a better trade-off between expressivity and learnability (Dehaene 2009).
[3] Staal (2006) also offers an overview of some of these factors.
[4] In many senses, the current approach follows Netz's (1999) idea of a 'cognitive history' (Netz also opposes the social constructivist (relativist) perspective of, e.g., Steven Shapin).
[5] However, a word of caution seems to be called for here: it must also be stressed that, as with virtually all technologies, there is always a trade-off involved. Precisely because they are a powerful technology, the use of formal languages also entails certain risks and dangers. In other words, it does not come 'for free', and things *can* go wrong (some such cases will be discussed in Chapter 3).

In what follows, I present an overview of the book, chapter by chapter. The aim is to provide a bird's-eye view of the overall purposes of the project, and of how its different elements all tie together.

Part I is composed of three chapters, and focuses predominantly on the history and philosophy of formal languages. It departs from the conviction that the conception of formal languages as mathematical objects only 'scratches the surface' of their actual cognitive impact.[6] Therefore, a new reconceptualization of formal languages is required in order to dig deeper. As already suggested, I propose to view formal languages and formalisms as a *cognitive technology*, i.e., as specific devices that may enhance and even modify the reasoning and cognitive processes of human agents in which they are involved. For this purpose, I discuss the following topics: two meanings of 'formal' that seem particularly relevant for the conception of formal languages as *formal*; the very concept of a formal language; the history of the development of this technology, which is thus presented as a *cultural product*; and reasons why one should or should not use formal languages and formalisms (specifically, but not exclusively, when doing logic).

Chapter 1 presents two senses of the term 'formal' that seem particularly relevant for the attribution of a formal character to formal languages and formalisms. Insofar as 'formal language' is not simply a set phrase, it is important to investigate more closely what is *formal* about formal languages and formalisms in order to improve our conceptual understanding of them. The two notions discussed are the formal as *de-semantification* and the formal as *computable*, and they remain central throughout the book.

Chapter 2 lays down the conceptual grounds for the whole investigation. It first presents a discussion of the very concept of a *language* (essentially adopting an evolutionary perspective), and analyses a series of crucial distinctions: spoken v. written languages, natural v. artificial languages, languages as practices v. languages as objects. In the second part, I discuss in which sense(s) formal languages are *languages* and in which sense(s) formal languages are *formal*. Then, I present an outline of the conceptualization of formal languages as a cognitive technology, which will guide the analysis throughout.

Chapter 3 presents a condensed history of the development of this technology. Focusing exclusively on the current stage of these developments (which we have no reason to believe have come to completion!)

---

[6] But, of course, the fact that formal languages *are* mathematical objects remains a crucial aspect of their uses in logic, in particular for metalogical investigations (see Chapter 3).

obscures the fact that the formal languages that we now have at our disposal are the product of very long and complex historical processes. Essentially, the development of formal languages is a specific chapter in the development of mathematical notations more generally, which in turn is a specific chapter in the development of writing systems. For the most part, the development of formalisms (including what we could describe as proto-formalisms) was driven by the search for more efficient tools for calculation/computation.

Chapter 3 also surveys the main views defended so far concerning the rationale for using formal languages when doing logic (by Leibniz, Frege, Wittgenstein, Carnap, etc.). I argue that none of them gets to the bottom of the phenomenon, even though there is some truth in each of them. From these proposals, three main functions for formal languages emerge: as expressive devices, as iconic devices, and as calculating devices. In the second part of the chapter, I argue that there are costs, risks, and limitations that must not be overlooked when using formal languages (for logic and other kinds of inquiry); it is, as always, a matter of trade-offs. This is important, as the essentially positive picture of the impact of using formal languages and formalisms presented here must not blind us to the fact that there are downsides too.

Part II is the truly empirically informed part of the investigation, where I offer extensive empirical support for the hypothesis of formal languages and formalisms as cognitive technologies. As a first step, I present an outline of the spontaneous reasoning patterns in humans, so as to be able to clarify in which ways formal languages represent a device that enhances, complements, modifies, and in some circumstances *corrects* these spontaneous patterns. It will also prove useful to review the recent empirical results on the neuroscience of reading, given the trivial but nevertheless often forgotten observation that formal languages are *written* languages. Neural mechanisms involved in reading processes can ultimately offer us insight into why formal languages and formalisms can make us reason in ways that we would otherwise not be able to (or with much difficulty); I describe this effect as the 'debiasing effect' of reasoning with formal languages.

Chapter 4 surveys some familiar results of research in psychology of reasoning, which strongly suggest that our spontaneous reasoning patterns are nothing like the precepts of 'logical reasoning' as traditionally construed (as also surveyed in, e.g., Evans 2002). I discuss in particular what has been described (Stanovich 2003) as a 'fundamental computational bias of human cognition', namely the systematic tendency to bring prior beliefs to bear when reasoning and solving problems. I also argue for a pluralistic conception of human rationality, one which makes room for different canons of reasoning in different life situations.

Chapter 5 presents an examination of formal languages from the point of view of *extended cognition* (Clark 2008; Menary 2010b). The chapter begins with a discussion of the neuroscience of reading and writing, essentially following the results reported in Dehaene 2009. I then turn to the concept of extended cognition and apply it to formal languages and formalisms. I examine the results reported in Stenning 2002 on how different formal languages (discursive v. diagrammatic) have an impact on a student's process of learning logic (in an institutional setting). Equally important are the results by Landy and Goldstone (2007a; 2007b; 2009) suggesting that working with formalisms crucially relies on sensorimotor processing. Finally, I claim that formal languages not only increase our computational power, but actually allow us to 'run a different software' when reasoning, as it were;[7] in this sense, my analysis of formal languages from the point of view of extended cognition falls squarely within Sutton's (2010) 'second-wave extended mind' framework and Menary's (2007a) 'cognitive integration' conception of extended cognition.

Chapter 6 examines the crucial concept of 'de-semantification', as introduced by Krämer (2003). The concept is also discussed in Chapter 1, but now I return to it against the background of the experimental results presented in previous chapters, and connect it with experimental research on the phenomenon of *semantic activation* (i.e., the cognitive processing taking place when an agent hears or reads words). I claim, however, that the concept of de-semantification by itself is not sufficient to account for the processes involved in reasoning with formal languages (or formalisms, more generally). A closely related concept is then introduced, namely the concept of 're-semantification'; it concerns the possibility of applying a given formalism, which is developed against a specific background, to a *different* problem, phenomenon, or framework. In such cases, rather than being considered as 'meaningless', the formalism is given a new meaning, but the debiasing effect of de-semantification (to be discussed in Chapter 7) also occurs.

Chapter 7 explains in more detail what I call the 'debiasing effect' of reasoning with formal languages and formalisms. It begins with a philosophical, schematic discussion of the very concept of formalization, which will pave the way towards an explanation of how de-semantification and sensorimotor manipulation of the notation are combined for the debiasing

---

[7] The reference to software here should be understood as metaphorical, and thus not as an endorsement of computational theories of the mind.

effect of reasoning with formal languages and formalisms to come about. However, it is important to notice that the debiasing effect is particularly prominent in cases of *applications* of formalisms, i.e., when logical or mathematical apparatuses are employed for the investigation of particular (extra-logical) phenomena – that is, when formalisms are *tools* rather than *objects* of investigation in themselves. Because it allows us to counter the 'computational bias' of systematically bringing prior beliefs into the reasoning process, reasoning with formal languages and formalisms in fact increases the chances of obtaining surprising results, i.e., results that go beyond previously held beliefs. This, I think, is the real magic, the *generosity*[8] of formal languages. I also discuss the work of Houdé and collaborators on inhibition training (e.g., Houdé and Tzourio-Mazoyer 2003), which represents a very different approach to debiasing, and draw some implications from the debiasing account of formal languages and formalisms for the currently popular dual-process model of human cognition.

I am well aware that the methodological approach adopted here is somewhat unconventional. There is still some opposition to the idea that philosophical analysis should be informed by empirical data (on human cognition or otherwise). Similarly, the relevance of historical investigations for systematic issues remains controversial. Now, if there are reservations towards the philosophical relevance of these two points of view taken in isolation, they are likely to intensify towards a philosophical analysis integrating *both*. I believe, however, not only that a wide range of philosophical problems can benefit from such an integrative approach, but in fact that they cannot be satisfactorily treated unless historical and empirical elements are taken into account. To elaborate on this point, in the concluding chapter, I discuss in more detail the methodological choices made for this investigation, and some of their implications in view of the results obtained.

---

[8] In the sense of d'Alembert's famous saying, 'Algebra is generous; it often gives us more than what is asked of it.'

I

Two notions of formality

Uses of the term 'formal' are ubiquitous in philosophical discussions, in particular (though not exclusively) in discussions pertaining to logic. But even a cursory examination quickly reveals that the term seems to be used equivocally and often in imprecise ways; arguably, it is essentially a blanket term covering a wide range of related but distinct concepts and phenomena. As a result, any analysis where the concept of the formal is supposed to play a prominent role seems to require a preliminary examination of the concept as such, and in particular a specification of which of its many variations will be relevant for the enterprise.

Perhaps the best known example of this methodologically commendable approach is MacFarlane's 'What Does it Mean to Say that Logic is Formal?' (2000), which presents a taxonomy of different senses of the formal in order to explore the (presumed) formal nature of logic. In his investigation, MacFarlane has a clear agenda, namely that of formulating demarcating criteria for logic on the basis of the notion(s) of the formal he discusses, which leads him to distinguish the different variations he presents in two main groups: 'decoy' and legitimate notions of the formal.[1] But, clearly, the classification is relative to MacFarlane's specific demarcating goals – something which is perhaps not made sufficiently clear by him. Many of his so-called 'decoy' notions are historically very significant, but their significance is buried under the 'decoy' label. For this reason, his taxonomy as a whole may not be entirely suitable against the background of different agendas.

I have presented in Dutilh Novaes 2011a an alternative taxonomy of different notions of the formal, primarily (though not exclusively) as pertaining to logic; the goal there is to analyse the notion of the formal as such, without the pressure of external goals. My proposed taxonomy has some overlap with MacFarlane's, but, rather than the 'decoy–legitimate'

---

[1] The term 'decoy' is MacFarlane's own terminology, but the term 'legitimate' is my interpretation of what he seems to have in mind.

dichotomy, its two main clusters are: the formal as pertaining to forms (five variations) and the formal as pertaining to rules (three variations). As pertaining to forms, the formal is essentially related to what something *is*; as pertaining to rules, the formal is essentially related to what something (or someone) *does*. Moreover, the opposite of the term 'formal' as pertaining to rules is (typically) the term 'informal', whereas the opposite of the same term as pertaining to forms is (typically) the term 'material'. So, clearly, there are important differences between the two clusters, which should arguably be captured by any proposed taxonomy.

In the present investigation, by contrast, I do have a rather specific agenda, namely that of exploring the cognitive impact of using formal languages and formalisms more generally when reasoning. It then becomes vital to raise the question of what is *formal* about formal languages (that is, if 'formal languages' is not merely a set phrase, which I do not believe it is) and formalisms, and in what sense their 'formality' impacts the cognitive processes involved. For this end, not all of the eight variations of the formal I identified in Dutilh Novaes 2011b will turn out to be equally relevant. In fact, I will argue that the two senses of the formal relevant for the cognitive approach adopted here are what I refer to as the formal as *de-semantification* and the formal as *computable*. Indeed, it must be stressed that, although intimately related, these two variations must be kept apart (incidentally, something that MacFarlane (2000) fails to do), and not only for conceptual but also for cognitive reasons, pertaining to the different layers of how using formal languages affects reasoning processes.

In this chapter, I offer a conceptual analysis of these two notions (their cognitive implications will be analysed in later chapters), paying attention in particular to their historical emergence within logic and its philosophy. (For the broader picture, in particular for the connections between these two variations and the other variations, see Dutilh Novaes 2011a.)

1.1 THE FORMAL AS DE-SEMANTIFICATION

I identified in Dutilh Novaes 2011b three variations of the formal corresponding to the idea of abstraction from subject-matter and content: the formal as topic-neutrality, the formal as abstraction from intentional content, and the formal as de-semantification.[2] The last corresponds to the

---

[2] This is roughly MacFarlane's 'syntactic formality' (2000: sect. 2.1), one of his 'decoy' variations. I find his terminology misleading here, as a syntactic approach does not necessarily take a stance concerning the meaningfulness or lack thereof of the language in question.

familiar idea that the abstraction in question concerns abstraction from *all meaning whatsoever*, i.e., that it amounts to what we could refer to as a process *de-semantification*[3] of some portions of (written) language. In this sense, to be purely formal amounts to viewing symbols as blueprints (inscriptions) with no meaning at all – as pure mathematical objects and thus no longer as 'signs' properly speaking. Indeed, when one speaks of 'formal systems', 'formal theories', etc., de-semantification is typically an important component.[4]

This variation of the formal is a fairly recent addition to the pool of different meanings – at any rate, if compared to the millennia-old notion of the formal as schematic (Dutilh Novaes 2011b). In effect, I have never come across any pre-nineteenth-century association of symbols being treated as non-symbols (i.e., as inscriptions without meaning) to the term 'formal'. In fact, in medieval theories of supposition (thirteenth and fourteenth centuries), the significative, meaningful kinds of supposition (personal and material) were taken to be *formal* precisely insofar as they were *significative*, because signification was associated to the notion of *form*. In contrast, when terms were considered in their materiality only, thus not in their meaningfulness, then *material* supposition occurred.[5] So, interestingly, we have exact opposite uses of the terminology in medieval theories of supposition and in the modern conception of the formal as 'meaningless'.

Historically, this variation of the notion of the formal has its roots in the nineteenth century and was later consolidated within Hilbert's programme on the foundation of mathematics in the 1920s.[6] Already in the *Grundlagen der Arithmetik* (1884) and later in 'Über formale Theorien der Arithmetik'

---

[3] I owe the term to S. Krämer (2003). See Chapter 6 for further details.
[4] Nevertheless, there is a crucial step from 'formal objects' understood as meaningless inscriptions to the inductive generation of a potentially infinite collection thereof by means of recursive rules. The latter aspect pertains to the notion of the formal as computable, to be discussed in the next section. I owe this point to Wilfried Sieg. A more detailed philosophical (as opposed to historical) analysis of the concept of de-semantification will be offered in Chapter 6, alongside a cognitive–empirical analysis of the concept.
[5] See my account of medieval theories of supposition (Dutilh Novaes 2011a). Notice that there are many other uses of the term 'formal' that have been important in the history of philosophy at one point or another, but which have by now disappeared from the (analytic) philosopher's terminological toolkit. Uses of 'formal' closer to the metaphysical origin of the terminology, in particular in connection with the Aristotelian notion of formal causes, were abundant in the Aristotelian tradition, but they are now mostly of historical interest. So, even though my main interest here is the history of the different variations of the formal that are now still pervasive in philosophical circles, my focus on these variations should not be interpreted as ignorance of these uses which are no longer current.
[6] 'To Hilbert is due . . . the emphasis that strict formalization of a theory involves the total abstraction from the meaning, the result being called a *formal system* or *formalism* (or sometimes a *formal theory* or *formal mathematics*)' (Kleene 1951: 61). But notice that Hilbert's 'formalistic' approach is fundamentally different from that of other 'formalists' (see below).

(1885) Frege criticizes this approach, in particular as defended by his colleague in Jena, Carl Johannes Thomae.[7] Responding to Frege's criticism, in the second edition (1890 – the first edition had been published in 1880) of his textbook *Elementare Theorie der analytischen Funktionen einer komplexen Veränderlichen*, Thomae writes:

For the *formal* conception, arithmetic is a game with *signs which one may call empty*; by this one wants to say that (in the game of calculation) they have no other content than that which has been attributed to them concerning their behavior with respect to certain rules of combination (rules of the game).[8] (emphasis added)

The gist of the formalist position defended by Thomae and, to some extent, by his contemporary H. E. Heine, seems to be a plea for ontological freedom: mathematical symbols need not be thought of as picking out (mathematical) objects and properties, and thus mathematical propositions do not state (mathematical) facts. (The ontological implications of de-semantification will be discussed in Chapter 6.) But, as is well known, Frege then went on to demolish this particular 'formalist' conception of mathematics; see Weir 2011.

Later, with Hilbert's programme in the 1920s, a variation of this sense of the term 'formal' became pervasive, especially once Hilbert's approach was described as 'formalist'.[9] However, it must be stressed that Hilbert's views are significantly more sophisticated than the idea of 'treating mathematical symbols as meaningless', so Hilbert's programme should not in any way be viewed merely as a continuation of Thomae's 'formal conception' of arithmetic.[10] The gist of the Hilbertian notion of the formal is aptly captured in the following passage by Bernays:

The typically mathematical character of the theory of provability reveals itself especially clearly, through the role of the logical *symbolism*. The symbolism is here the means for carrying out the formal abstraction. The transition from the point of view of logical content to the formal one takes place when one ignores the original meaning of the logical symbols and makes the symbols themselves representatives of formal objects and connections.

[7] In nineteenth-century English mathematics, the idea of treating mathematical symbols as 'meaningless' was also present, in particular with Peacock and Babbage (I owe this point to W. Hodges).
[8] Translation taken from Thomae's biography, available at www-groups.dcs.st-and.ac.uk/~history/Biographies/Thomae.html. The term 'formal' as used here is most probably reminiscent of the Kantian idea of the formal as abstraction from content I discussed in Dutilh Novaes 2011b.
[9] The term 'formalism' was initially introduced as a pejorative characterization of Hilbert's position by Brouwer. Formalism remains to this day a respectable position in the philosophy of mathematics, albeit with a relatively small number of proponents (see Weir 2011: sect. 7).
[10] See Chapters 3 and 6 and Weir 2011 for further details.

For example, if the hypothetical relation
'If A then B'
is represented symbolically by

$$A \rightarrow B$$

Then the transition to the formal standpoint consists in abstracting from all meaning of the symbol $\rightarrow$ and taking the connection by means of the 'sign' $\rightarrow$ itself as the object to be considered. To be sure one has here a specialization in terms of figures instead of the original specialization in terms of content; this, however, is harmless insofar as it is easily recognized as an accidental feature. Mathematical thought uses the symbolic figure to carry out the formal abstraction. (Bernays 1930: 248)

A key term in this passage is 'formal abstraction': abstraction is of course a crucial concept in mathematics, and, from an Aristotelian perspective, abstraction typically corresponds to abstraction of form from matter. Here, however, abstraction corresponds to ignoring specifically the *meaning* (content) of signs; as the inverted commas used by Bernays indicate, 'signs' thus viewed are no longer signs once they become the objects to be considered in themselves, precisely because they are de-semanticized. Significantly, for our purposes, both the process of abstraction *and* the result obtained (a 'purely formal system') are both qualified as 'formal'. In a similar vein, Carnap writes at the very beginning of *The Logical Syntax of Language*:

A theory, a rule, a definition, or the like is to be called *formal* when no reference is made in it either to the meaning of the symbols (for example, the words) or to the sense of the expressions (e.g., the sentences), but simply and solely to the kinds and order of the symbols from which the expressions are constructed. (Carnap 1934: 1)

On the one hand, the formal as de-semantification may be seen as a radical variation of the general 'abstraction from content' Kantian motto; in effect, Kantian ideas had considerable influence over Hilbert and Bernays.[11] But, on the other hand, there is a sense in which the formal as de-semantification can also be seen as stemming from the formal as schematic (Dutilh Novaes 2011b), as described in the following passage by Tarski:

---

[11] Bernays (1930) mentions Kant several times. See also MacFarlane 2000: 142 n. 10 for additional historical connections between the programme of 'rigourization' of mathematics and a general Kantian background, and Zach 2003: sect. 2.1 on the influence of Kant over Hilbert and Bernays. In fact, it is fair to say that nineteenth-century mathematics and philosophy in Germany in general were heavily influenced by Kant, albeit in different ways.

Already at an earlier stage in the development of the deductive method we were, in the construction of a mathematical discipline, supposed to disregard the meanings of all expressions specific to this discipline, and we were to behave as if the places of these expressions were taken by variables void of any independent meaning. But, at least, to the logical concepts we were permitted to ascribe their customary meanings ... Now, however, the meanings of all expressions encountered in the given discipline are to be disregarded without exception, and we are supposed to behave in the task of constructing a deductive theory as if its sentences were configurations of signs void of any content. (Tarski 1959: 134)

To be sure, the adjective 'formal' used in this sense also qualifies a certain methodology, that is to say a specific manner of proceeding – the 'mechanical' manipulation of symbols – which will be discussed in more detail in the next section. But it is crucial to distinguish the formal as de-semantification from the formal as computable, as it is possible to proceed 'formally', i.e., mechanically, even without the preliminary step of 'formal abstraction' from meaning. Similarly, it is quite a conceptual step from treating signs as meaningless to the purely computational manipulation of signs. Now, one way to illustrate the independence of the two notions is to provide examples of proponents of one of the two approaches but not of the other. So, arguably, while Thomae's conception represents the formal as de-semantification but not the formal as computable,[12] Frege's conception of the foundations of mathematics (in particular with respect to the role of his 'ideography') arguably relies on the formal as computable but not on the formal as de-semantification. In fact, he makes the latter patently clear in his debate with Hilbert (Blanchette 2007). Let us thus examine the formal as computable in more detail.

1.2 THE FORMAL AS COMPUTABLE

Uses of the term 'formal' as synonymous with 'computable' are extremely pervasive in logic and philosophy of logic of the last eight decades, the period which witnessed a revolution in digital computing. In this section, I first present a tentative archaeology of uses of the term 'formal' in the sense of computable or algorithmic. Following that, I discuss the very concept of computable in order to attain a better grasp of what it means to be formal in this particular sense.

[12] Frege points out that Thomae and Heine 'do not set out an account of the syntax and proof theory which is remotely adequate as an account of the mathematics with which they deal' (Weir 2011: sect. 2).

1.2.1 Uses of formal as computable

Uses of the term 'formal' as synonymous with 'computable' are extremely pervasive in the logical developments of the last eight decades; for example, a 'formal system' is typically defined as a recursively axiomatized theory (but not necessarily one whose symbols are treated as 'meaningless'). But, such uses appear to be a relatively recent phenomenon in philosophy and logic: unlike the variations of the formal as pertaining to forms, which have an old and distinguished philosophical pedigree (Dutilh Novaes 2011b), 'formal' in the sense of computable seems to have become widely used only in the twentieth century. Moreover, it can be argued that, according to a more precise use of the terminology, the extension of 'computable' is actually a proper sub-class of the extension of 'formal' in this very sense, as there are systems with which one operates 'formally', i.e., according to explicitly stated rules, but which do not display full computability in that they are not recursively axiomatizable. Nevertheless, uses of the term 'formal' as synonymous with 'computable' have become current practice.[13] More generally, the idea is that to operate 'formally' is to follow 'mechanically' the instructions contained in a calculus, i.e., a well-defined notational system with clear rules of formation and transformation – an idea made particularly conspicuous in Carnap's *The Logical Syntax of Language*.

In the 1930s, interest in the concept of computability emerged against the background of Hilbert's programme, in particular with respect to the so-called 'decision problem' (*Entscheidungsproblem*). The title of Turing's (1936) seminal paper, introducing the notion of a Turing machine, is 'On Computable Numbers, with an Application to the *Entscheidungsproblem*'.[14] In effect, in the 1930s, the concept of computability and effective calculability was perhaps the most widely discussed topic in logic and the foundation of mathematics, to a great extent as an aftermath of the blow inflicted by Gödel's incompleteness theorems to Hilbert's programme.[15] Uses of 'formal' in the sense of computable in this period were probably directly influenced by uses of the same terminology in connection

---

[13] For example, the widespread view according to which second-order logic is not 'logic' properly speaking because its set of validities is not recursively enumerable.

[14] 'By the *Entscheidungsproblem* of a system of symbolic logic is here understood the problem to find an effective method by which, given any expression Q in the notation of the system, it can be determined whether or not Q is provable in the system' (Church 1936: 41). Sieg (2008) offers a detailed account of how the modern technical notion of computability emerged from attempts to solve the decision problem.

[15] 'In the 1930s, well before there were computers, various mathematicians from around the world invented precise, independent definitions of what it means to be computable. Alonzo Church defined the Lambda calculus, Kurt Gödel defined Recursive functions, Stephen Kleene defined Formal

with the formalistic programme. Seemingly, these early uses of the term 'formal' in the sense of computable essentially derive from the formal as de-semantification.

Notice, however, that Turing himself does not use the term 'formal' in this sense in his 1936 paper. But many of the researchers involved in the development of the concept of computability – such as Kleene and Gödel – ostensively used the term 'formal' as synonymous with computable, in particular in the phrase 'formal system'. Curry remarks that the notion of a formal system

is quite a different notion from that of a postulate system, as naively conceived a half century ago. In the older conception a mathematical theory consisted of a set of postulates and their logical consequences. The trouble with this idea is that no one knows exactly what a logical consequence is ... In the modern conception this vague and subjective notion is replaced by the objective one of derivability according to explicitly stated rules. (Curry 1957: 1)

These explicitly stated rules should allow for 'mechanical' implementation, and it is in this very sense that the system is said to be formal. Similarly, commenting on the problem of decidability for mathematical theorems, Gödel says:

The first part of the problem has been solved in a perfectly satisfactory way, the solution consisting in the so-called 'formalization' of mathematics, which means that a *perfectly precise language* has been invented, by which it is possible to express any mathematical proposition by a formula. Some of these formulas are taken as axioms, and then certain rules of inference are laid down which allow one to pass from the axioms to new formulas and thus to deduce more and more propositions, the outstanding feature of the rules of inference being that they are purely *formal*, i.e. refer only to the *outward structure* of the formulas, not to their meanings, *so that* they could be applied by someone who knew nothing about mathematics, or by a machine. (Gödel 1995: 3–45; emphasis added)

This passage by Gödel illustrates the connection between the 'formal' view of symbols as pure shapes and the concept of the formal as computable: once the total abstraction from meaning is undertaken, the different transformations and manipulations of symbols may become a purely mechanical endeavour. Nevertheless, it is in fact a dramatic step (something that is not made sufficiently explicit by Gödel's phrase 'so that') from the idea of treating (mathematical or otherwise) symbols as meaningless to the rule-governed inductive generation of a *potentially infinite* collection of

systems, Markov defined what became known as Markov algorithms, Emil Post and Alan Turing defined abstract machines now known as Post machines and Turing machines' (Immerman 2004: sect. 1).

('meaningless') formulae, which could then be used to express mathematical propositions, as described by Gödel. This step could only be taken once the notion of computability was sufficiently mature, which was clearly not the case prior to the 1930s developments.[16]

It is also important to notice that algorithmic procedures are used on two distinct levels within a formal system, both of which are mentioned in Gödel's passage. On one level, the formulae of the language are inductively generated by means of the application of its rules of syntax to its basic terms (rules of formation), generating a potentially infinite collection of formulae. On another level, the rules of inference (rules of transformation) defined for the formal system can be 'mechanically' applied to the formulae in order to establish what follows from what within the system, and in particular what its theorems are. These are the formulae obtained by the application of the rules of transformation to axioms, in first instance, and then to the formulae obtained from the axioms, in a procedure that can be iterated as many times as one wishes. So a formal system is 'formal' both in the sense of how its language is generated and in the sense of the transformations performed within the system.[17]

One might expect that Frege would have used the term 'formal' in the sense of computable. However, Frege was arguably not particularly concerned with the *mechanical* move from premises to conclusion as such; rather, his main concern was that every inferential step be made thoroughly explicit, i.e., that there be no gaps in the expression of an argument (see, for example, the preface of the *Begriffsschrift* or his critique of Dedekind's lack of explicitness concerning inferential steps in the *Grundgesetze*). For Frege, logic is an epistemological tool allowing for the production of new knowledge, which takes place when a person effects the inferential passage from judgments (premises) to new judgments (conclusion), the passage itself being a judgment as well. Thus, Frege's goal was *epistemological clarity*, not 'mechanical reasoning' as such, although the requirement that no hidden contentual considerations be incorporated into the application of rules can be read as a demand for a 'mechanical' application of rules.[18] Thus,

---

[16] Although Gödel is not usually seen as having made significant contributions to the topic of computability, he exerted a significant influence on some of the founders of the field such as Church, Kleene, and Rosser (see Zach 2006).

[17] Notice that, in this sense, a system can be 'formal' even if it is not decidable, i.e., even if it cannot prove or disprove all the formulae of its underlying language.

[18] See Sieg 1994: 74–5. Moreover, 'I here want to consider two views, both of which bear the name "formal theory". I shall agree with the first; the second I shall attempt to confute [Thomae's]. The first has it that all arithmetic propositions can be derived from definitions alone using purely logical means, and consequently that they also must be derived in this way' (Frege 1885: 94; translation, p. 141). By

one may say that the formal as computable was not an end as such for Frege, but rather a means towards epistemic transparency.

Another likely precursor of uses of 'formal' in the sense of computable was Leibniz, given the crucial role of both the notion of calculability and of the terminology of 'form' and 'formal' in his investigations. However, as far as I could establish, he does not use the term 'formal' in the distinctive sense of computable or calculable. He often speaks of calculating procedures and of forms (both in connection with logic), but so far I have not encountered any explicit occurrence of the term 'formal' as such as qualifying calculating procedures. Here is a passage where both the notions of form and calculation occur, but not the adjective 'formal':

> Thus I assert that all truths can be demonstrated about things expressible in this language with the addition of new concepts not yet expressed in it – all such truths, I say, can be demonstrated by a calculus alone, or *by merely manipulating the characters according to a certain form*, without any effort of the imagination or the mind, in a word just as it is done in arithmetic and algebra. (Leibniz 1982: 195, emphasis added)

Thus, it is true that the association of computations and calculations with forms is already to be found in Leibniz, and therefore that the meaning of 'formal' as computable is in a sense already present there, but only in a germinal state. But if the notion of the formal as computable is not to be found (in any case not fully developed) in Leibniz, then it is probably not to be found in any other author prior to the 1930s.[19] In short, it can be asserted with a reasonable degree of certainty that, in logic and philosophy, uses of the term 'formal' in the sense of computable became consolidated only in the 1930s, in first instance in connection with its uses in the context of Hilbert's formalistic programme.[20]

1.2.2 *What does it mean to say that something is computable?*

The question of what it means to say that something is formal can thus (partially) be given a straightforward answer by reference to the notion of computability – effectively calculability – as this is indeed one important

---

'purely logical means', Frege surely meant the rules put forward in the *Begriffsschrift*, and in this sense it can be said that Frege was already relying on a proto-form of the formal as computable. Naturally, this is also a clear formulation of Frege's logicism.

[19] To my knowledge, it is in particular not to be found in the tradition of the 'algebra of logic' stemming from Boole's work.

[20] But, notice that, for my overall goals, nothing of significance hinges on this historical claim.

sense in which the term is used. But, of course, this is only a relative answer, and a rather uninformative one if not accompanied by a discussion of what it means to be *computable*, i.e., by a proper understanding of the concept of computable as 'not requiring any insight or ingenuity'.

In this section, I offer a brief discussion of the informal concept of computability relying of course on the technical definitions formulated by Turing, Church, and Gödel, but focusing essentially on understanding its *conceptual* nature as it is represented in these different but equivalent definitions.[21] A proper understanding of the concept of being computable as in 'not requiring any insight or ingenuity' will be essential for the present investigation; it will be a central aspect of the analysis of the cognitive impact of uses of formal languages to be presented subsequently.

In the 1930s, the different ongoing projects of these three authors – Turing, Church, and Gödel – all converged into the same focal point, namely the notion of computability (Sieg 2006: sect. 1). What must be kept in mind is, however, that prior to the formulation of technical notions such as that of a Turing machine, general recursive functions, and lambda-definability, the notion of computability was originally an *informal* notion.[22]

But what, then, are the core elements of the informal notion of being computable? Historically, theoretical discussions of this notion came to occupy a central position only in the twentieth century; prior to that, few authors had given it much thought as such. Of course, the term 'algorithm' had been employed in mathematics at least since the Middle Ages,[23] as roughly synonymous with 'calculating procedure', but again with scarce conceptual analysis of the concept itself, let alone a precise technical

[21] It is often noted that the very fact that these rather different formal characterizations of computability – Turing machines, Church's lambda-definability, and Gödel's general recursive functions – are proved to be equivalent suggests that there is indeed some deep, 'core' notion of computability which is correctly captured by these formal characterizations (see P. Smith 2009: chaps. 33–4). Other analyses of the notion of effective calculability (by Post and Markov, among others) have also been proved to be all equivalent. The convergence of these different definitions is usually seen as strong evidence for their correctness (see Copeland 1997: 'The thesis and its history').

[22] 'Thus, we see in the work of these three pioneers [Turing, Church, and Gödel] the absolute need for a *mathematically precise concept* corresponding to the *informal notion* of effective calculation or mechanical procedure' (Sieg 2006: 192; emphasis added). But notice that this does not mean that the gap between the informal notion of computability and its technical counterparts cannot be bridged, as many seem to think. Smith (2009: chap. 34) argues convincingly that a variation of Kreisel's 'squeezing argument' can be used here. I think that Turing's analysis is not only formally correct but also conceptually insightful, and the idea of a Turing machine truly captures the conceptual core of the notion of computability.

[23] The term 'algorithm' comes from a 'Latinization' of al-Khwārizmī, the name of the ninth-century Persian mathematician whose books on algebra and arithmetic were very influential (see section 3.1.1, below).

definition. Nevertheless, an examination of these pre-twentieth-century discussions of the notion of computability can be insightful in that they explore the conceptual dimensions which were then somewhat overshadowed by the subsequent technical developments.

The machine metaphor. A recurrent and to some extent illuminating explanation of the concept of effective calculation or computation relies on the notion of a *mechanical procedure*. But what is a *mechanical* procedure? In other words, how do *machines* in general operate? Well, the basic intuition seems to be that they implement previously determined actions in a sequence, step by step, in a completely predictable way, thus reaching the desired outcome.[24] The mechanism of a (well-functioning) clock, for example, goes from state to state by means of the previously defined (within the mechanism itself) operations of its parts. To calculate is thus in first instance an action undertaken by a human agent, but one who acts *as if she was a machine*, following predetermined patterns, such as those specified by means of an algorithm. To calculate or compute is to reason as machines would if they could reason.

Of course, now that computers are ubiquitous, the expression 'mechanical thinking' does not strike us as particularly absurd, but one must appreciate that this is a very recent development. Remember that Descartes considered all animals to be 'machine-like' precisely because, unlike humans, they do not have a soul and thus cannot think. We now seriously entertain the idea of thinking machines – as testified by the lively field of artificial intelligence – but until fairly recently the idea of mechanical thinking would at best be seen as a metaphor, as thinking was considered to be a prerogative of very specific kinds of creatures (chiefly human beings, but under certain ontologies also God and other immaterial thinking entities, e.g., angels).

There are actual implementations of 'thinking machines' prior to the twentieth-century development of computers, although they are quite rare and rudimentary. In particular, researchers working in the field of artificial intelligence often view the idiosyncratic Ramón Lull as the grandfather of the field.[25] Lull thought that he could use combinatorial machines as means

---

[24] In fact, these are a particular sort of machine, the deterministic sequential machines, but they are arguably the sort of machine underpinning the 'machine metaphor' in its most basic formulation.

[25] 'Many of the fundamental ideas in artificial intelligence have an ancient heritage. Some of the most fundamental, surely, are that thinking is a computational process, that computational processes involve combining symbols, that computation can be made mechanical, and that the mathematics of computation involves combinatorics. All of these ideas have their origin, so far as we know, in the work of an eccentric 13th century Spanish genius, Ramon Lull (1232–1316)' (Ford, Hayes, and Glymour 1998: 136).

to prove to infidels the existence of God, thus implying that one could construe rational arguments by means of machines.

In the seventeenth century, there was quite a lot of discussion on the possible application of calculation methods to produce new knowledge, i.e., to 'think mechanically', in particular with Leibniz.[26] The following description (in modern terminology, not that of Leibniz) of the gist of Leibniz's general approach to the matter is illuminating:

> Leibniz's project stands out for its emphasis on mechanical reasoning: a universal character is to come with algorithms for making and checking inferences. The motivation for this requirement emerges from his complaint about Descartes's *Rules for the direction of the mind*. Leibniz views them as a collection of vague precepts, requiring intellectual effort as well as ingenuity from the agents following the rules. A reasoning method, such as the universal character should provide, comes by contrast with rules that completely determine the actions of the agents. Neither insight nor intellectual effort is needed, as a mechanical thread of reasoning guides everyone who can perceive and manipulate concrete configurations of symbols. (Sieg 2008: 531)

Curiously, one of Descartes's criticisms of scholasticism and logic in general was precisely that logic sends our reason 'on holiday', as no special perspicuity is required for the application of logical reasoning.[27] Contrary to Descartes, Leibniz saw this very feature of logic as a major asset, allowing for greater epistemological reliability. But, of course, Leibniz's 'thinking machines' (which in theory would also operate on combinatorial principles, just as Lull's) never were realized. In contrast to Lull, for Leibniz, the idea of mechanical thinking remained essentially a metaphor, and the calculators he refers to are clearly meant to be humans: his famous motto '*Calculemus!*' (Let us calculate!) is an imperative addressed to people, not to machines.

Also, note that even Turing, when introducing the notion of (what we now call) a Turing machine in his 1936 paper, did not refer to machines properly speaking. The agent operating the 'machine', i.e., undertaking the actions determined by the table of instructions, is actually a human being, to whom he refers as the 'computer'. The human computer simply follows strictly the instructions contained in the table of instructions of a given Turing machine, thus not adding any insight of his/her own.[28]

---

[26] Notice, however, that Leibniz was far from being alone in the seventeenth century with his interest in artificial languages; quite a few others shared his interest (see Chapter 3; Maat 2004; Mugnai 2010).

[27] *Rules for the Direction of the Mind* (Descartes 1985: 1: 36, Rule X).

[28] Crucially, the table of instructions is an external element, and Turing's analysis contains an important component of *externalization* of reasoning which can be fruitfully viewed from the point of view of extended cognition (see Chapter 5).

Now, just as a Turing machine is (originally) not a machine properly speaking, as it is operated by a human, the description of a 'mechanical procedure' as a procedure where every step is completely determined by previous states does not in fact describe the operations of actual machines. As any owner of a car, a washing machine, or even a computer knows, machines often do unpredictable things, i.e., they act differently at different occasions even when the same buttons seem to be pushed: same input, different outputs. So, clearly, when speaking of a 'mechanical procedure', it is not so much actual machines that we have in mind but rather the *ideal* of a machine, i.e., the abstract machine which always behaves according to plan, that is according to the instructions contained in it. To illustrate this distinction, I now turn to Wittgenstein's *Philosophical Investigations*:

193. The machine as symbolizing its action: the action of a machine – I might say at first – seems to be there in it from the start. What does that mean? – If we know the machine, everything else, that is its movement, seems to be already completely determined.

We talk as if these parts could only move in this way, as if they could not do anything else. How is this – do we forget the possibility of their bending, breaking off, melting, and so on? Yes; in many cases we don't think of that at all. We use a machine, or the drawing of a machine, to symbolize a particular action of the machine. For instance, we give someone such a drawing and assume that he will derive the movements of the parts from it. (Just as we can give someone a number by telling him that it is the twenty-fifth in the series 1, 4, 9, 16 . . .)

'The machine's action seems to be in it from the start' means: we are inclined to compare the future movements of the machine in their definiteness to objects which are already lying in a drawer and which we then take out. – But we do not say this kind of thing when we are concerned with predicting the actual behaviour of a machine. Then we do not in general forget the possibility of a distortion of the parts and so on. – We *do* talk like that, however, when we are wondering at the way we can use a machine to symbolize a given way of moving – since it can also move in quite *different* ways.

But when we reflect that the machine could also have moved differently it may look as if the way it moves must be contained in the machine-as-symbol far more determinately than in the actual machine. As if it were not enough for the movements in question to be empirically determined in advance, but they had to be really – in a mysterious sense – already *present*. And it is quite true: the movement of the machine-as-symbol is predetermined in a different sense from that in which the movement of any given actual machine is predetermined.

Indeed, the 'given way of moving' described by Wittgenstein, corresponding to the machine-as-symbol, amounts precisely to the notion of 'formal' in its computational variation. But, of course, Wittgenstein also reminds us that actual machines do not (necessarily) behave 'mechanically', especially but not exclusively due to hardware limitations and constraints. Thus, when we speak of a mechanical way of proceeding, we in fact typically have in mind ideal machines, not real ones.

The intuitive notion of computability. A calculation/computation can be defined as the passage from one initial state (the premises, or more generally the initial state of information) to a final state (the conclusion, the result of the calculation), by means of successive 'small' passages from state to state. It is a successive transformation of enumerable, finite states, yielding a particular output from a given input.[29] Given that the idea is to obtain *some* output, a crucial element in the notion of calculation is that it be a *finite* procedure; it is a procedure that must come to a halt.[30] It can be represented as:

$$x_1 \Rightarrow x_2 \Rightarrow \ldots \Rightarrow x_n$$

The key point is defining the rules or instructions allowing for the passages from a state to the next state in such a way that any agent would effect the passage in exactly the same way,[31] i.e., obtaining the same X_j for any X_{j-1}, $j \leq n$. Seen this way, a computation has a few fundamental properties, such as: it is discrete, i.e., it is not a continuous process, rather it progresses through discrete, individual steps; it is a dynamic process, taking place in time and with a clear temporal direction; every step is completely determined,[32] i.e., the instructions are applied strictly, requiring no insight or ingenuity. An algorithm or effective calculation method is a procedure that determines

---

[29] 'An algorithm, we said, is a step-by-step procedure which can be fully specified in advance of being applied to any particular input. Every minimal step is "small" so that it is readily executable by a calculator with limited cognitive resources. The rules for moving from one step to the next are entirely determinate and self-contained. And an algorithmic procedure is to deliver its output after a finite number of computational steps' (Smith 2009: 320).

[30] This is indeed an essential aspect of the notion of computability, but one that is unanswerable for Turing machines: some Turing machines on certain inputs never halt. This is known as the halting problem. Typical ways in which a program (a computation) is non-terminating are cases of loops or of infinite progress. But this is not a flaw of the definition of a Turing machine; because it is intended to be perfectly general, it must assume that the tape where the symbols are written is infinitely long and that the time for the calculation is also infinite, so that it does not rule out functions as non-computable simply because the machine runs out of tape or time.

[31] Interestingly, the notion of an effective calculating procedure brings along a democratic component; anyone mastering the procedure can undertake it, whereas in the absence thereof, insight and ingenuity would be required. Krämer (2003) speaks of the distinction between 'knowing why' and 'knowing how': I do not need to understand the reasons why a given procedure works to be able to use it.

[32] Of course, there are now theories of non-deterministic computation. But here I will focus on the classical notion of computability, as it is arguably the important one in order to understand the concept of formal as computable.

univocally, by means of explicit instructions (which must be expressed by means of a finite number of symbols), what X_j is to be for any X_{j-1}, $j \leq n$.

Naturally, this description invokes the notion of a *function*, i.e., an operation taking arguments into uniquely determined values (at most one value, but possibly none). But as is well known, not all functions are effectively computable, and the famous Church thesis (Church 1936) is the claim that a function of positive integers is effectively computable only if it is recursive.[33]

After the formulation of Church's thesis, Turing proposed another (but equivalent, as we now know) analysis of the notion of computability, one in which the informal features of this notion are better represented than in Church's thesis: the concept of (what we now refer to as) Turing machines (he called them 'logical computing machines'). Turing's main accomplishment is to provide a precise formulation for the rather vague informal idea that a calculating procedure should involve no ingenuity or insight, i.e., that the instructions in question can be applied 'mechanically'. A Turing machine[34] is a state machine, constituted of an infinite one-dimensional tape divided into cells (it is infinite in only one direction) and a read–write head, which can move from cell to cell and can read the symbol contained in a cell or write a symbol in it. Each cell can contain one symbol, either '0' or '1'. Its actions are determined by a table of transition rules, specifying for each state j and for each symbol (either 0 or 1) what the read–write head should do next. Each table of transition rules defines a different Turing machine, while the 'hardware' (the tape with cells and the read–write head) remains the same. A transition rule is a 4-tuple:

$$< State_j, \; Symbol, \; State_{j+1}, \; Action >$$

The possible actions are: write a symbol in current cell (thus overwriting the symbol currently in it); move one cell to the right; move one cell to the left. Once the next action is not completely determined by the state and the symbol in a given cell – i.e., if according to the table of instructions either no action or more than one action is prescribed for a given state *n* and symbol *s* – then the procedure comes to a halt, and the computation comes to an

---

[33] Gödel introduced the concept of general recursive functions in his 1934 Princeton lectures after it had been shown by Ackermann that there are effectively computable functions which are not primitive recursive (Wang 1990: 15). So, clearly the notion of computability required a broader technical definition than just merely corresponding to the class of primitive recursive functions, hence Church's thesis.

[34] There are quite a few (equivalent) ways of defining a Turing machine. Here I follow the presentation of Barker-Plummer (1995).

end. Thus described, a Turing machine appears to be an exceedingly simple device, but the cornerstone of Turing's analysis is the proof that such a simple device can nevertheless calculate any recursive function of positive integers (Turing 1936).[35] The point is that, given the initial state j and the symbol written on it, the action to be undertaken by the machine and what the next state is going to be are completely determined by the table of transition rules, which in practice is the 'program' for the machine.

For our purposes, it is important to emphasize the crucial role of *concrete, external symbolic systems* at the very heart of the notion of computability. Generally, algorithms depend critically on their instantiation in actual portions of writing. Of course, once mastered, an algorithm (say, 'carrying' when undertaking sums within the familiar Hindu–Arabic numeral system) can be simulated mentally, but this in fact corresponds to an *internalization* of the external cognitive artefact (the procedure as defined in a given notation). While claiming that symbolic instantiation is a *necessary* (constitutive) condition for something to count as an algorithm or calculating procedure is an exceedingly strong claim, it is sufficient to notice that, in practice, human beings have developed calculating procedures in close connection with external notational systems (which, I will argue, says something very deep about the structure of human cognition).[36]

In effect, Turing's description of a Turing machine makes the connection between computing/calculating and what can be described as an *externalization* of reasoning even more patent: the tape with cells, the read–write head and the table of instructions are all external, physical devices involved in the calculating procedure – cognitive artefacts. From this point of view, the formal as computable is inherently related to the idea of cognitive extension/integration (to be discussed in Chapter 5). This also means that, when applied to the notion of *formal* languages, the formal as computable clearly suggests that an extended cognition perspective (such as the one to be adopted here) is particularly suitable to investigate the effects of using formal languages and formalisms when reasoning.

Another way of formulating roughly the same point is that the informal concept of computability is thus from the start motivated by inherently *cognitive* considerations: how do we exclude appeal to 'ingenuity and insight' from certain reasoning processes?[37] In Chapter 5, I will argue that

---

[35] See Smith 2009: chap. 30 for some examples.
[36] See Dutilh Novaes, 'Mathematical Reasoning and External Symbolic Systems' (forthcoming).
[37] By contrast, as we will see in Chapter 6, the avowed philosophical motivations for the move towards de-semantifications are typically related to considerations of ontological parsimony. Attention to the *cognitive* implications of de-semantification has been scarce thus far.

notations which incite sensorimotor manipulations greatly enhance the possibility of blocking the interference of ingenuity and insight (thus blocking tacit external assumptions and reasoning shortcuts). A passage by Whitehead illustrates this idea particularly well:

> By relieving the brain of all unnecessary work, a good notation sets it free to concentrate on more advanced problems, and in effect increases the mental power of the race . . . [I]n mathematics, granted that we are giving any serious attention to mathematical ideas, the symbolism is invariably an immense simplification . . . *[B]y the aid of symbolism, we can make transitions in reasoning almost mechanically by the eye*, which otherwise would call into play the higher faculties of the brain. It is a profoundly erroneous truism . . . that we should cultivate the habit of thinking of what we are doing. The precise opposite is the case. Civilization advances by extending the number of important operations which we can perform without thinking about them. (Whitehead 1911: 59–61; emphasis added)

Indeed, the less we 'think about what we are doing', the less likely it is that ingenuity or insight will be called upon. This is a theme that will remain current to the end of this study.

1.3 CONCLUSION

In this chapter, I explored the two notions of the formal which seem most relevant for the concept of formal languages; what I have elsewhere (Dutilh Novaes 2011b) called the formal as *de-semantification* and the formal as *computable*. Their cognitive implications will be discussed in more detail in Chapters 5, 6, and 7, but the goal of this chapter has been to argue that these two notions are historically rich, philosophically refined, and thoroughly embedded in debates in the philosophy of mathematics and on the notion of computability. The main upshot is thus again that integrating the philosophical and the cognitive perspectives is a natural development given the underlying commonalities between the two perspectives.

CHAPTER 2

On the very idea of a formal language

Within the project of acquiring a better grasp of the cognitive impact of using formal languages and formalisms, it is not sufficient to resort to the familiar idea – in fact, the mathematical definition – of a formal language as a language composed of a basic lexicon of symbols and strict formation rules.[1] For the purposes of philosophical analysis, a more promising approach seems to be to reflect on what is *formal* about formal languages, and in what sense(s) formal languages are *languages*. Beyond the apparently merely verbal nature of these questions, my claim is that addressing these issues will also shed light on the general topic of the present enterprise. The discussion of two senses of the formal in Chapter 1 will provide the starting point for the question concerning the formality of formal languages. Similarly, a brief discussion of what it means for something to be a *language* will prove useful, so as to establish in what sense(s) formal languages are indeed languages, and in what sense(s) they are not like 'ordinary' languages.

2.1 'LANGUAGE' AND LANGUAGES

2.1.1 *What is a language?*

From the start, it is not obvious whether raising an utterly general question such as 'what is a language?' can lead to fruitful discussion. The

---

[1] This conception of formal languages is to be traced back to Carnap's *The Logical Syntax of Language* (see Chapter 3). Here is a fairly standard way of presenting formal languages: 'In mathematics, a formal language is normally defined by an alphabet and formation rules. The alphabet of a formal language is a set of symbols on which this language is built. Some of the symbols in an alphabet may have a special meaning. The formation rules specify which strings of symbols count as well-formed. The well-formed strings of symbols are also called words, expressions, formulas, or terms. The formation rules are usually recursive. Some rules postulate that such and such expressions belong to the language in question. Some other rules establish how to build well-formed expressions from other expressions belonging to the language. It is assumed that nothing else is a well-formed expression' (Sacharov 1999).

concept of language is so ubiquitous and wide-ranging that to inquire into the 'essence of language' is likely to be viewed as a 'grammatical joke' (Wittgenstein 1953: § 111). Indeed, any putative set of necessary and sufficient conditions governing our uses of the term 'language' is bound to cover only a small portion of such uses. The problem is not only that the term is very often used in a metaphorical sense (which is also the case); more importantly, it is unclear whether there could even be agreement on where the language/non-language boundary exactly lies.

Nevertheless, I submit that some observations on what it means for something to be a language can be offered so as to pave the way towards a better understanding of the phenomenon of formal languages. Indeed, I suggest that the semantics of the term 'language' can be understood as based on a prototypical core of phenomena that are incontestably viewed as languages, and that other uses of the term refer to phenomena varying in different degrees from the core phenomena (similar to the semantics of colour terms, for example). To be clear, I am not claiming that the core phenomena provide necessary conditions for something to *be* a language, but rather that other phenomena are referred to as 'languages' insofar as they bear some resemblance (in varying degrees) to the core phenomena.

But what are these core phenomena? Here I want to propose something that is not entirely uncontroversial, but which seems like a reasonable assumption/working hypothesis: the core phenomena that guide our understanding of the concept of a language are the spoken languages used by humans. These, I claim, are the prototypical uses of the term 'language'. Of course, positing that only what humans use as means for oral communication should count as 'language' is rather stipulative and not innocent. In effect, in debates on the extent to which non-human animals are linguistic animals, the assumption that only our own system of oral communication can be rightly described as a language simply begs the question. Nevertheless, notice again that the claim that human spoken languages count as the prototypical cases of uses of the terminology is not an essentialist claim; rather, it is a semantic assumption, one not extensively argued for here, but meant to be the starting point for further discussion.

Granting the hypothesis that human spoken languages can play the role of paradigmatic uses of the term 'language', we still have not made much progress unless we are able to identify some very general features of these phenomena, which in turn warrant different uses of the term to refer to other phenomena that display these features to a greater or lesser extent. Now, a list of the core features of human spoken languages is again a highly controversial matter. Nevertheless, there seem to be four features in human spoken languages that are widely regarded as central: (1) speech; (2) semantics; (3) syntax; (4) communicative function.[2] I will discuss each of these in turn, but let me note that, following the general methodological guideline favouring empirically informed philosophical analysis adopted here, I will mostly be relying on empirically established facts about human languages and their counterparts in non-human animal communication.[3] None of these features is here aprioristically taken to be a key feature of human spoken languages; rather, they emerge from extensive empirical research. Moreover, the comparative approach with non-human animals and the emphasis on evolutionary accounts of language are fruitful in that they compel us to reflect on what exactly is specific (if anything at all) to human spoken languages.

(1) *Speech.* A crucial characteristic of these ubiquitous human signalling systems, spoken languages,[4] is indeed that they rely on our vocal capabilities. Among animals (including humans), communication goes through a variety of different modalities, ranging from the scent of secreted substances (as in bees, ants, wasps) and the sight of bodily gestures (as in dogs, wolves, rabbits), to communication through touch. But speech/sound is indeed one of the most common means of communication for animals, second perhaps only to scent. Humans are not alone in possessing an elaborate vocal system that allows them to articulate a multitude of sounds. We do have 'a descended larynx, hyoid and tongue root which enable us to produce a greater diversity of formant frequencies patterns' (Fitch 2005: 198), but this is no longer thought to be unique in humans. Moreover, it is important to notice that animals in different phylogenetic lineages have also developed sophisticated vocal systems (birds, cetaceans), which suggests that the development of such systems is not exclusively driven by the purported evolutionary advantage of the emergence of specific vocal signalling systems such as human spoken languages.[5] So vocalization remains a fundamental characteristic of human communication but is not, as such, exclusive to

---

[2] For example, Fitch (2005) analyses the evolutionary emergence of the first three features in the human species by means of comparisons with signalling systems in other species, and notes that communication is a very plausible candidate concerning the evolutionary advantage that human spoken languages could represent to the members of the species displaying this trait.

[3] I will be drawing mostly on Fitch 2005, as he presents a comprehensive review of the literature on the evolution of language.

[4] In effect, every known human population engages in practices that we uncontroversially refer to as 'speaking a language'. Writing, on the other hand, is much less widespread a phenomenon (see also below).

[5] Interestingly, the great apes, our closest still extant phylogenetic relatives, have vocal systems that are quite rudimentary when compared to animals belonging to different clades.

humans. The capability of vocal imitation is not exclusive to humans either (the obvious example being parrots), but it is certainly a necessary condition for humans to have developed their spoken languages.

In the opposite direction, it is also important to notice that there are many human signalling systems that do not rely on vocalization as their expressive modality; in other words, to reduce human languages to spoken languages would be excessively restrictive. Of course, we have 'natural' signalling systems going through scents and gestures just as the majority of non-human animals. But beyond these, humans now regularly communicate through two media that are essentially visual: written languages and sign languages. The extent to which these two kinds of signalling systems are derived from human spoken languages will be discussed shortly, but it is clear that one's position on this matter will be decisive in order to determine how central speech is within human languages. In any case, it is clear that human languages, broadly construed, at this time, go beyond speech.[6]

(2) *Semantics*. Obviously, human spoken languages are typically *meaningful*, but this truism is, as it stands, not very helpful; what it means for a piece of language to be meaningful is a thorny issue, one which philosophers of language, linguists, and cognitive scientists continue to debate vigorously. As a first approximation, we can of course note that spoken words typically stand for 'things', i.e., they bear a semiotic relation to entities other than themselves. (Recall the Lilliputians in *Gulliver's Travels* who decided to do away with words by carrying objects around instead, in order to communicate with others.) This is a fundamental fact, and at least to some extent an important step in a child's process of language acquisition. But, as can be argued on philosophical (e.g., Wittgenstein in the *Philosophical Investigations*) and empirical grounds (as outlined by the work of Michael Tomasello and collaborators), merely naming objects does not seem to be our primary use of spoken languages.

More generally, meaningfulness can be conceived of in terms of the relation of representation that portions of language bear with respect to portions of the world, or, alternatively (more generally), in terms of the impact that the occurrence of a given portion of language may have in the speaker's interactions with other speakers and with the world. The

---

[6] On the basis of the observation that the areas of the brain thought to be involved in language and in motor skills overlap significantly, it has been hypothesized that human languages first developed as gesture languages rather than as vocal languages (Fitch 2005: 219–20). While compelling, this hypothesis is now viewed as lacking sufficient empirical corroboration to be considered as a serious contender among theories on the evolution of human languages.

first approach can be described as a 'representational' account of meaning; the second approach could be described as a 'usage-based' account of meaning;[7] Loeffler (2009) refers to this approach as 'practice-based theories of meaning'.

Having signals that stand for things (i.e., what we could describe as systematic 'referential calls') is again not something that is unique to human beings (Fitch 2005: 204–6). Yet it is significant that no animal is known to have as wide a vocabulary as humans do. Upon training, apes can learn as many as 200 signs, but the amount of training involved is considerable; by contrast, human children typically have a vocabulary of around 6,000 words as first-graders, which they acquire predominantly from exposure alone (no intensive training required). Indeed, the neural mechanisms that seem to be involved in the establishment of relations of meaning between signs (be they pictures, written or spoken words, etc.) and things are quite similar in humans and other non-humans primates, but the gap in the amount of vocabulary is very significant. Moreover, as pointed out by Tomasello (2003), the signals used by non-human animals are typically understood by all their conspecifics, whereas human spoken languages are understood only by members of the same linguistic community.[8]

A remarkable feature of how the human brain processes such meaning relations (and which is to some extent present in non-human primates) is that different areas of the brain seem to be in charge of different kinds of objects. So, a given area is typically activated when a participant is exposed to a sign corresponding to an animal, for example, to the point that brain lesion in that area may lead to the specific loss of all spoken and written knowledge of animals, to the extent that some patients cannot even recognize animals in pictures anymore (Dehaene 2009: 112).

In other words, there seems to be a strong neurological component in the phenomenon of word meaning (in the 'representational' sense). The conventionality and arbitrariness of language cease once a given speaker has learned a given word. In Dehaene's (2009: 113) words, 'we cannot help seeing an animal behind the word "lion", and find it impossible to read the word "giraffe" as a verb – each of these words has become solidly attached, by temporal lobe connections, to the many dispersed neurons that give it

---

[7] The reference here would be to 'usage-based theories of language acquisition' (Tomasello 2003). The latter, however, are more encompassing in that they also include a claim of the primacy of lexical competence over grammatical competence; here, the contrast is with representational theories of meaning more specifically.

[8] However, there is evidence of 'cultural variation' in communicating practices of cetaceans (Rendell and Whitehead 2001).

meaning'.[9] Nevertheless, Dehaene (2009: 113) himself acknowledges that, although much progress has been made, 'how meaning is actually coded in the cortex remains a frustrating issue'.

Furthermore, while animals do seem to be able to form concatenations of referential calls, these are typically very simple and short, not on a par with the complex combinations of words that form sentences in human spoken languages. Such combinations, according to the principle of compositionality (Pagin and Westerståhl 2010), give rise to propositional, complex meaning. Again, how exactly this cognitive process takes place remains an elusive aspect of the human linguistic abilities.

However, a crucial divide between animals and humans with respect to meaningfulness and semantics is perhaps best explained in terms of a 'usage-based' conception of meaning. Tomasello (2003) and collaborators have famously claimed that non-human animals, including apes, lack a 'theory of mind', and this is why their calls are not intentionally referential on the part of signallers. Essentially, the idea is that no animal other than humans seems to be able to deploy the Gricean mechanism of intentions and recognition of intentions which (arguably) underscores human verbal communication. For example, according to Tomasello, one of the earliest pre-verbal forms of communication in human infants is the 'Look! This is interesting!' intervention, 'as when an infant points to a passing bird for its mother and squeals with glee' (Tomasello 2008); this, he claims, is unique to humans. So, if Tomasello is right, the main difference between human and non-human animal communication seems to pertain not so much to the ability to associate portions of language to portions of the world, but rather to the *uses* that they each make of languages, i.e., what they seek to obtain both in terms of reactions from other members of the species and in terms of the effects produced in the surrounding environment.

(3) *Syntax.* The extent to which syntax is a key element of human spoken languages is perhaps the most controversial of the four aspects discussed here. Nobody would deny that human spoken languages display a high level of systematicity in how words are combined to form complex expressions. In fact, productivity in human spoken languages depends crucially on the possibility of such combinations. Rather, the bone of contention pertains to the level of structural complexity that can be attributed to *actual* uses of spoken language by humans (as opposed to the competence level of what humans *could* do if they wanted to), and to the centrality of grammatical

---

[9] See Chapters 5 and 6 for more on the 'neuroscience of meaning'.

competence (as opposed to lexical competence) in the mastery of a human spoken language (Tomasello 2003: chap. 1).

Partisans of the Chomskyan hypothesis of the existence of a 'universal grammar' famously claimed that recursion is a universally present feature in all human spoken languages, and furthermore that it is unique to human languages (Hauser, Chomsky, and Fitch 2002); it represents the core of human linguistic competence. What they mean by 'recursion' is essentially the fact that complex phrases can be embedded (nested) into longer phrases: 'structures can be iteratively embedded in similar structures to generate progressively more complex structures, e.g., phrases within phrases' (Fitch 2005: 195). Given the purported universality of this feature in all human spoken languages, it has been claimed that humans have an 'innate' ability for recursion thus understood, which moreover would have been the result of an evolutionary process of natural selection (Pinker and Bloom 1990).

A human spoken language not displaying these characteristics would then presumably refute such claims of universality, and this is why the purported discovery of such a language, namely the language spoken by the Amazonian Pirahã Indians (Everett 2008), generated so much stir. However, whether the Pirahã language does or does not display recursion thus understood remains a controversial matter (Nevins, Rodrigues, and Pesetsky 2009; Everett 2009).

In practice, though, in most *oral* linguistic exchanges of any human language (even those viewed as highly 'complex'), nesting of phrases into other phrases rarely goes beyond one level of nesting, and speakers typically have great difficulties understanding the meaning of phrases with several iterations of nesting (Fitz 2009: chap. 6).[10] This can of course be attributed to the competence–performance gap, but the fact that in actual oral exchanges our presumed capacity for nesting is scarcely called upon raises doubts as to how crucial recursion really is for human linguistic abilities. Nesting/embedding of sentences is indeed much more widely present in *written* linguistic contexts, and one wonders whether theorists are not in fact projecting properties of written languages into their analyses of human spoken languages (on the differences between written and spoken languages, see section 2.1.2, below). In fact, some have argued (and convincingly, to my mind) that linguistics as a discipline suffers from a chronic 'written language bias' (Linell 2005).[11]

---

[10] There are exceptions, of course, e.g., oral poetry and children's rhymes.
[11] 'the models and methods used for studying spoken language, and language in general, are largely those inherited from times when the goals were those of standardising and exploring written language' (Linell 2005: 31).

Be that as it may, the fact of there being underlying rules and systematicity guiding the formation of complex expressions from simple linguistic items is undoubtedly a feature of human spoken languages.[12] While other animals are apparently able to produce and understand short sequences of discrete signs (Fitch 2005: 206), and while there has been increasing interest in the sophisticated 'syntax' of songbirds' vocalizations (van Heijningen, de Visser, Zuidema, and ten Cate 2009), to date no similar level of complexity to that of even fairly simple human oral linguistic exchanges has been observed in non-human animals.

Hence, even if recursion (in the sense of the capacity to formulate nested phrases) may not be as central to human spoken languages as has been claimed, these languages do seem to rely on rules of concatenation and composition. Besides the *semantic* conception of compositionality (the meaning of a complex expression is a function of the meaning of its parts and how they are put together), equally significant is the *syntactic* conception of compositionality: complex phrases are formed by means of the grammatical composition of simpler phrases and single expressions. Philosophers often postulate that compositionality is a necessary feature of human languages, but again this seems to be a hypothesis to be investigated empirically rather than a matter to be settled on conceptual grounds alone.

In this respect, an exciting field of research on the evolution of human languages concerns uses of computer models to investigate how languages themselves, rather than the language-producing individuals, evolve (Kirby, Smith, and Brighton 2004). The evolution in question is not biological, phylogenetic evolution; rather, it is the *cultural* evolution of every new generation of speakers who learn a language as children introducing slight modifications to the language they are exposed to initially. '[L]anguage learners induce a grammar based on the output of past language learners, and because this output is limited and imperfect, the induced grammar will often be slightly different from that of the model, resulting in language change' (Fitch 2005: 208). Indeed, language learning appears to be one of the main motors behind language change (Niyogi and Berwick 1997). Now, what is remarkable in such computer simulations is that the different stages in the evolution of a language (not necessarily an actual language, but often a 'toy example' designed for the purposes of computational modelling) tend to converge towards languages displaying a high level of regularity, which

---

[12] For the purposes of the present investigation, nothing hinges on the presence or absence of recursion as a universal feature in all human spoken languages. I tend to sympathize with the view that recursion is *not* a universal feature and that grammatical competence is not the core of human linguistic abilities, thus siding with Tomasello (2003), but it is immaterial for the present investigation.

are thus easier to learn than completely unsystematic communication systems (also referred to as holistic communication systems).

What these results seem to show is that compositionality, as understood by philosophers (Pagin and Westerståhl 2010), is perhaps not a *necessary* condition for something to count as a language (as postulated by philosophers) but rather a feature that suits the cognitive constraints of the agents in question. In other words, it is not that a language cannot *but* be highly compositional; rather, the iterated process of learning a language through generations selects and produces languages which are highly compositional because they are easier to learn and to operate with from the point of view of human agents. Such a non-aprioristic argument for the significance of compositionality in human languages is to my mind much more compelling than simply positing on conceptual grounds that compositionality is a necessary feature of human languages. Hence, the conclusion seems to be that syntax and grammar are important features in human spoken languages possibly in virtue of the constrained cognitive make-up of the members of the human species. But grammar may well be an epiphenomenal (albeit widely present) feature of human languages, arising through processes of grammaticalization (Tomasello 2003: 5), which in fact are mostly driven by meaning-related factors.

(4) *Communicative function*. Looking at human spoken languages from an evolutionary point of view inevitably leads to the question of the evolutionary *advantages* (if any) that possessing signalling systems such as human spoken languages must have represented for those individuals within the species who happened to develop such systems. Given all the 'remarkable' things that human languages allow us to accomplish nowadays (literature, philosophy, science), it is all too easy to jump to the conclusion that possessing such languages must have represented an evolutionary advantage for those individuals also during the relevant evolutionary period. But, of course, it would be entirely unwarranted to infer that the development of spoken languages among humans corresponded to a selected-for trait in the crucial period of phylogenic evolution of the human species solely on the basis of how useful they are to us now.

In other words, that the capacity of using spoken languages did represent an evolutionary advantage and thus that it triggered the well-known mechanisms of natural selection among our ancestors cannot simply be assumed. It may well be that the particular trait of being able to speak is predominantly a *spandrel*: it may not have represented any particular evolutionary advantage at the time of the phenotypic appearance of the trait. Instead, it may have arisen as a by-product of other traits that were themselves adaptations, such as larger brains and developed cognitive capacities, in particular with respect to planning (Steedman 2002).

There are, however, a few robust hypotheses on how the emergence of human spoken languages may have been a direct product of selection (natural or otherwise), which go well beyond the level of being 'just-so stories'. For our purposes, what matters is that at least two of them directly involve the communicative function that we tend to view as a fundamental aspect of human uses of spoken languages.

One theory for the evolution of language which does *not* rely on the purported communicative function of human spoken languages is based on the concept of sexual selection. It relies on 'an implicit assumption that complex language could increase mating success directly, by making the speaker attractive to potential mates' (Fitch 2005: 211). Fitch then goes on to argue convincingly against this hypothesis; one of his arguments is that human language considered as a trait does not display the same features of traits in other species which clearly have been selected for on the basis of sexual selection (the quintessential example being the tail of the peacock). One such feature is the early maturation of the language behaviour in humans, whereas traits that are selected for on the basis of sexual selection in other species typically only appear around puberty.

Fitch's own preferred explanation for the evolutionary emergence of human languages relies on the concept of *kin selection*. The idea is that if the main evolutionary issue is not the survival of a particular individual but rather the survival of its genes, an individual who is seemingly altruistic towards its own close kin is effectively protecting its own genetic material, or in any case genetic material very close to its own. With respect to language in particular, given that 'communicating kin often share each other's genetic best interests, kin selection will favor an individual (especially a parent) who can increase a relative's (particularly, offspring) survival by communicating information about food or predators' (Fitch 2005: 213). From this point of view, individuals who could speak and thus communicate more extensively made better parents in that their offspring were subsequently better prepared to survive on their own. These in turn could transmit the same knowledge received (and whatever additional knowledge amassed during their lifetimes) to their own offspring, establishing the cumulative process of cultural transmission.[13]

---

[13] It has been proposed that the altriciality of human beings, i.e., the fact that they are born very immature and require extensive nurturing for a long period of time, must have played a crucial role in the evolution of language. 'Hominid infants appear to have evolved a specific tendency to use

The kin-selection hypothesis is attractive in many respects, for example in that it explains the onset of 'honest', truthful communication; the mechanism would only work if parents would convey truthful information to their offspring. But this hypothesis is not the only one available; another popular hypothesis, which is, however, not discussed in Fitch 2005, concerns the communicative advantages of spoken languages for the purpose of coordinating joint activities involving several individuals. The quintessential example would be that of hunting; coordination of several hunters may yield a better pay-off to each of them individually than if they each went hunting on their own. But, for coordination to be effective, a sophisticated signalling system would be required, and the existence of complex communicative systems such as the spoken languages of humans would presumably enhance stable family and group structures. However, unlike the kin-selection hypothesis, the social coordination hypothesis has difficulties explaining the emergence of truthful communication. In effect, well-known game-theoretical dilemmas such as the stag hunt suggest that cooperation does not necessarily yield the best pay-off; defection may often be a more attractive option, and untruthful communication would be a form of defection. Moreover, it has been suggested that individuals who focus on small game which can be hunted individually are overall better fed than those involved in the practice of large game-hunting. More generally, models of language evolution based on coordination and cooperation among (non-kin) individuals must respond to such challenges adequately; compare the attempt of Demichelis and Weibull (2008), among others.

Some authors have emphasized the connections between language and planning (Stenning and van Lambalgen 2008: chap. 6; Steedman 2002). The main idea would be that the development of language is related to the development of the capacity for planning in humans; both systems would display structural similarities (e.g., nesting), and there is also neurological evidence supporting this connection. Planning here must not necessarily be understood as *social* planning, i.e., as planning for activities involving several individuals (in which case communication is, of course, crucial). Nevertheless, the planning approach coupled with the inherently social nature of the human species quite naturally suggests the idea that language, planning, and coordination are all part of the same story. Stenning and van Lambalgen stress in particular the altriciality of human beings (i.e., the

elaborate vocalization as a means of soliciting long-term investment from caregivers' (Oller and Griebel 2006: 293). There are, however, different ways in which altriciality may have been an important factor for language evolution (see more on altriciality below).

developmental immaturity of human offspring upon birth) as a possible factor in the establishment of more complex cooperative social settings, which would in turn create a background where truthful, sophisticated communication becomes crucial.[14]

In summary, a few plausible accounts of how the evolutionary emergence of human spoken languages would have followed classical mechanisms of selection and of increased fitness for the individuals in question (and for the species) have been proposed. According to these accounts, it is precisely the enhanced communicative possibilities afforded by a more complex signalling system that presumably allowed for increased fitness. However, the matter is not entirely settled, so we cannot as yet (and possibly never will) be sure that the communicative advantages of human spoken languages did in effect increase fitness in the appropriate environment.

Furthermore, it is important to bear in mind that complex communication is not a feature exclusive to humans. It is well known that several non-human animals have signalling systems that are highly effective for the purposes of communication and coordination among members of the same species (Gibson 2010). The already mentioned cases of ants and bees are well-known examples, but sophisticated systems of communication have also been observed in whales, dolphins, monkeys, and apes. Still, that human languages as used now play a significant communicative role is something that can hardly be denied, and which seems in any case to be a key feature of how and why humans use spoken languages *now*.

What I hope to have accomplished with these observations is to establish that there are good conceptual and empirical reasons to view these four aspects as crucial to our understanding of what is to count as a language: speech, meaningfulness, syntax, and communicative function. In section 2.2.1 I discuss the extent to which each of these aspects is present in formal languages.

2.1.2 *Written v. spoken languages*

I have taken as a starting point above the idea that human spoken languages represent the paradigm of what is to count as language, even though humans have also developed languages that use different media (most

---

[14] 'There are good general reasons for believing that the social co-ordination required for the establishment and preservation of language conventions requires a highly cooperative setting such as that altriciality provides. For example, Davidson's arguments for the "principle of charity" in interpretation [52] provide just such reasons. If we were really so focused in our earliest social dealings on whether we are being cheated or lied to, it seems unlikely that human communication would ever have got off the ground' (Stenning and van Lambalgen 2008: 148).

notably writing and sign-languages). There are (at least) two possible interpretations of this view: (1) it is essentially correct, given that these other human languages are ultimately all derived from spoken languages; (2) it must be qualified, as these other languages must not be seen as mere secondary counterparts of spoken languages in different media. The second reaction would be based on the conviction that writing in particular (but possibly also sign-languages) is essentially autonomous as a signalling system vis-à-vis speech. In this section I defend this position, and suggest that failure to appreciate the fundamental dissimilarities between speech and writing is a fatal flaw in much theorizing about language, ranging from philosophy of language and linguistics to formal semantics. In particular, it might be a conceptual mistake to place speech and writing under the common heading of 'natural language'; I take issue both with the purported commonality between the two forms implied by the unique designation and with the attribution of 'naturalness' (see also below).

Although still essentially under-appreciated by the scientific community as a whole, the speech v. writing divide has been extensively discussed by a number of theorists who have gone on to offer a reconceptualization of writing – in particular, Harris (1986; 1995; 2009), Olson (1994), Krämer (2003), and Derrida (1998).[15] Common to these authors is the idea that writing should not be seen as 'nothing more than an ingenious technical device for representing spoken languages, the latter being the primary vehicle of human communication'.[16] One of the advantages of such a reconceptualization is the possibility of a better understanding of forms of writing which do not fit into the traditional picture, such as musical writing and mathematical notation, among others. Formal languages are one such form, hence the importance of a more detailed discussion of writing in the present context.

The received, standard view on writing can be referred to as 'phono-graphic' (Krämer 2003: 519) or 'phonoptic' (Harris 1995: xii): the purpose of writing is exclusively that of making sound (spoken languages) *visible*, i.e., to picture to the eye what the spoken word sounds like to the ear. This conception clearly reserves a secondary place to writing vis-à-vis speech within human communication in general. Central to the standard conception is the purported superiority of writing systems which closely follow the phonetic aspects of speech, alphabetic or syllabaric writing systems in particular, given that the ultimate goal of writing is after all to represent

---

[15] The present analysis will draw predominantly on Harris and Krämer.
[16] G. Lussu, *La lettera uccide* (Viterbo: Nuovi Equilibri, 1999, p. 11), as quoted in Harris 1995: xi.

speech. Indeed, as noted by Harris (1995: 2), in many languages the term used to designate a person unable to read and write makes explicit reference to an alphabet (French: *analphabète*; Portuguese: *analfabeto/a*), the implication being that literacy is to be equated with mastery of *alphabetic* writing.

Harris and others have extensively discussed a number of shortcomings in the traditional phonographic/phonoptic conception of writing, which I will not review here. For our purposes, what is crucial is to notice that such a conception of writing cannot accommodate the formal languages of logic and other formalisms, which are from the start *not* conceived as having explicit counterparts in speech. In turn, failure to emphasize the apparently platitudinous fact that formal languages are written languages prevents us from fully understanding the role(s) they play in the practices of logicians.

Let me just mention one important element that the phonographic/phonoptic conception of writing does not do justice to: the history of the development of writing. Writing is not a universally widespread technology: many current societies still do not rely on writing to any significant extent. Moreover, historically, the vast majority of linguistic populations never developed anything like written languages autonomously. Writing does appear to have emerged independently in a few different locations (Mesopotamia, China, and Mexico), but in all other cases it is imported technology from elsewhere. In particular, as shown by the work of Schmandt-Besserat (1996), the very early development of writing in Mesopotamia had no connection whatsoever with the idea of representing speech. Rather, proto-forms of writing were used for *counting*, more specifically for the bookkeeping of goods and production surplus. The first developments in this direction date back to 7500 BC, and for almost 5,000 years (although there were developments concerning the surfaces on which the signs were inscribed), writing remained exclusively used for accounting purposes. It was only around 2700–2600 BC that writing was first put to use for something other than bookkeeping, namely for funerary purposes: to write the names of deceased kings on objects to be deposited in tombs. After that, the development towards more comprehensive writing systems was rather speedy, and, by 2000 BC, writing was used in historical, religious, legal, scholarly, and literary texts in the Sumerian courts.[17]

True enough, already at an early stage (around 2500 BC), short religious inscriptions (prayers) already displayed a structure similar to the syntax of spoken languages, with subjects and verbs. But cuneiform is a pictographic

[17] For reasons of space, I will not comment on the development of other independent writing traditions, but the general points made here essentially hold of them as well.

system of writing, so the idea of representing sounds by symbols (rather than content or meaning) was at that point not present. Of course, it can be argued (and has been argued) that the progress of writing naturally leads to alphabetic writing systems, which are thus 'superior' to other forms of writing (Harris criticizes this position extensively). Still, what the history of writing shows is in fact a wide diversity of writing systems, many of which were not in any way intended to represent speech. Even the concept of a linear progression towards phonetic writing does not withstand further scrutiny, and the fact that there are presently writing systems that are perfectly functional and widely used (e.g., Chinese writing systems) suggests that nothing essential hinges on the conception that phonetic writing is the 'true' kind of writing.

It might be thought that the findings on the cognitive science of writing, as reported by Dehaene (2009), in fact contradict the claim that writing should not be conceived as having as its primary goal that of graphically representing speech. The history of writing, for example, can be seen as gradually converging to alphabetic forms of writing precisely because these systems would be more suitable for the human cognitive make-up. In a sense, there would have been a process of cultural evolution with respect to writing, not fundamentally different from the process through which languages become more and more systematic (as discussed in section 3.1.1). It is true that alphabetic writing systems seem vastly to simplify the learning of writing skills, and that, typically, writing becomes a more democratic technology once such simplified, phonetic writing systems are introduced.

But Dehaene himself presents several reasons why the purely phonetic conception of writing is at best too limited, but possibly simply wrong. For starters, he develops at length the idea of 'two routes for reading' (Dehaene 2009: 38–49), one going from visual input to sound representation and another going from visual input directly to meaning (to be discussed in Chapter 5). They are both activated when humans read, and even though the first route is the first one to be developed ontogenetically, i.e., when a child is learning to read and write, the competent adult reader seems to rely predominantly on the letters-to-meaning route when reading. Secondly, in many languages, English in particular, there are countless words that are homophonic, i.e., which sound exactly the same but are actually different words – a fact which is then mirrored in different spellings. 'I' and 'eye', 'won' and 'one', and many more such pairs of words, sound identical when pronounced but are spelled differently, indicating their different meanings. So, at least in such languages, writing cannot be purely phonetic on pain of leading to excessive ambiguity and lack of clarity. Dehaene refers to this phenomenon as 'the impossible dream of transparent spelling'.

In other words, while a phonetic approach to writing appears to be instrumental in the early stages, becoming a proficient reader–writer consists exactly in going beyond the phonetic approach and making full use of the specificities of writing that have no counterpart in spoken languages.

However, what is perhaps surprising in much of the theorizing about language in different fields is that while it relies on a conception of writing that views it as secondary and derivative with respect to speech, we also observe what can be described as 'the written language bias' (Linell 2005). In a sense, this is not so surprising: if writing is thought to be a perfect representation of speech, then the assumption is that language theorists can focus on just one of the two media and get the other one 'for free'. Now, since writing is obviously easier to work with than speech, given its permanence, it is not surprising that theorists would focus on writing for convenience.[18]

But, once we part with the assumption of faithful correspondence between speech and writing to view different forms of writing as essentially autonomous linguistic systems, it becomes clear that the projection of many of the features of written languages into spoken languages may be unwarranted (e.g., the centrality of recursion in spoken languages discussed above). By the same token, many of the features of spoken language that have no counterpart in writing tend to be neglected by such studies. Even today, when much lip service is being paid to the need to study spoken languages as such, i.e., involving elements such as prosody, focus, and gestural/facial activity, the conceptual apparatus is still essentially dominated by concepts developed for the study of written languages. The conclusion is thus that the assumption of parallelism between spoken and written languages is hindering our proper understanding of both, in their specificities and peculiarities.

In effect, one of the peculiarities of writing that has no obvious counterpart in speech is what Krämer (2003) describes as its fundamental *iconic* nature. Writing takes place in space, typically on two-dimensional surfaces, yet the idea that writing is a representation of speech imposes on the former the linear one-dimensional structure of the latter. In this way, the inherent iconicity of writing is obscured and misunderstood. Iconic devices such as the use of italics and other font styles, tables of contents, capital letters, footnotes, etc., rely on the pictorial nature of writing for their expressive

---

[18] Linell (2005) offers other historical and sociological reasons for the prevalence of the written language bias.

functions, but these features typically do not receive sufficient attention from theorists. Now, the iconicity of writing is again a crucial element to be borne in mind if we are to attain a deeper understanding of uses of formal languages, as will be discussed in Chapter 3.

So, for convenience, let me briefly list the main differences between speech and writing as widely acknowledged in the literature, drawing on Linell 2005: chap. 3. First of all, speech obviously pertains to sound and hearing while writing pertains to vision; by the same token, we may want to say that speech essentially pertains to (one-dimensional) *time* while writing essentially pertains to (two-dimensional) *space*. Relatedly, speech is typically ephemeral (except if recording takes place) while writing entails a form of permanence. Moreover, speech interactions typically rely on a wide range of contextual elements for meaningfulness to arise – in particular (but not exclusively) the bodily gestures of the speakers – whereas writing often has to be 'more explicit' and self-contained given the presence of fewer contextual elements. Indeed, typically (though not always) speech concerns a multi-agent situation of interaction while writing does not presuppose a real-time, present interlocutor. From a cognitive point of view, and as already noted, there is a fundamental difference between the amount of learning and training required to master speech as opposed to master writing. While the wide majority of human beings learn to speak just by being adequately exposed to a language, writing requires intensive training and only very rarely does it come about just by means of exposure. Indeed, every healthy human being, having had the proper exposure, is able to speak, but writing remains a skill that a significant portion of human beings do not master. It is also often remarked that in speech people use language freely, while in writing they feel compelled to follow grammatical rules more closely.

Of course, a number of caveats to these oppositions can be raised, and in particular the impact of new technologies has to some extent blurred the traditional boundaries between speech and writing (e.g., real-time chatting has many, but not all, of the characteristics typically attributed to spoken linguistic interactions). Still, it seems more appropriate to view writing as an independent signalling system, not directly derived from spoken language, if we are to understand what both speech and writing are about in their specific peculiarities.

2.1.3 Natural v. artificial languages

A dichotomy that must also be looked into in the present context is the natural v. artificial dichotomy, in particular with respect to languages.

'Natural language' has become a set phrase, used to designate the languages that human beings 'ordinarily' speak and write, usually in contrast with so-called 'artificial' languages such as formal languages. But there seem to be (at least) two basic problems with this account: (1) the natural v. artificial divide with respect to languages is at best a difference of degrees, with several intermediate steps, thus not a clear-cut distinction; (2) the natural v. artificial dichotomy and its close relatives (nature v. nurture; nature v. culture; innate v. acquired) is conceptually problematic, in spite of its pervasiveness. According to an influential and roughly 'Aristotelian' conception of nature – as in, for example, Aristotle's *Physics*, book 2, chap. 1 (1984) – 'nature' is defined by opposition to what pertains to humans and the products of human actions; 'nature' is whatever is left once we isolate what is human made (often loosely described as 'culture'). But, obviously, the human v. nature opposition disregards the fact that human beings are themselves *part of* nature, and that human interventions and actions are always constrained by the same basic physical, chemical, and biological laws of the universe. At the same time, humans are not the only animals systematically to modify their 'natural' environment (e.g., the dams built by beavers). Now, while a sustained critique of the natural v. artificial dichotomy would be out of place in the present context, here I offer some considerations in order to substantiate the claim that thinking about the differences between formal languages and so-called 'natural' languages in terms of this dichotomy is a fundamentally misguided approach.

Let us start by assuming for a moment that there might be an essential difference between so-called natural languages and so-called artificial languages. Does this translate into a sharp boundary between the two? One criterion that is often appealed to in order to clarify this distinction concerns the amount of time involved in the development of a given language: while so-called natural languages evolve through a long span of time, typically involving a large number of people, presumably so-called artificial languages are the product of the deliberation of one individual in a very short period of time.

But this criterion disregards the fact that the appearance of a so-called artificial language typically relies on previous developments which themselves took place over long periods of time. As will be described in Chapter 3, formal languages (in their different variations) emerged within the extensive history of the development of mathematical notation, ranging over several centuries. Focusing on the single individual that proposes a particular systematization of a formal language at a particular time (say, Frege and his *Begriffsschrift*) obscures the fact that there is a wide range of previous developments that made this event possible at all.[19] Moreover, given that time is continuous, naturally a time-span criterion is bound to be prone to the usual vagueness phenomena: how much time exactly does it take for the development of a language to qualify as being natural (artificial)? What about borderline cases? Again, the distinction can at best be one of degrees.

One may also argue that a fundamental difference between so-called natural and so-called artificial languages is their range of applications. So-called natural languages are used in everyday life, while so-called artificial languages are typically restricted to specific contexts, most notably *scientific* contexts. Indeed, when first presenting his *Begriffsschrift*, Frege contrasts it with what he calls *Sprache des Lebens* (he does not refer to the natural v. artificial dichotomy). But, here again, the difference can only be one of degrees (Wang 1955). Indeed, the languages used in different disciplines such as, for example, sociology or biology are not fundamentally different from 'everyday' languages, except perhaps for the systematic use of technical terminology. In mathematics, there is wider use of specific symbolism, but typical mathematical proofs (i.e., outside the realm of mathematical logic) are still essentially written in 'plain' words; in fact, logicians themselves also make extensive use of 'plain' words in the articles they write.

One may think that there is a fundamental difference in cognitive ontogeny between so-called natural and so-called artificial languages; while the former arise 'naturally' in any community of human beings, the latter arise only in very specific cultural circumstances, typically related to a high level of scientific development. Now, if this is true at all (and to some extent it is), it only holds of *spoken* languages.[20] As we noted above, writing emerged independently in only a handful of places, and typically in the context of highly specific cultural and social circumstances. From this point of view, writing can be considered as 'natural' only insofar as it is a representation of spoken languages; but, given that this assumption is problematic in many respects (as argued above), it is unclear in what sense writing can be said to be 'natural' from an ontogenic point of view (similar considerations on the differences in learning to speak and learning to write could also be offered).

[19] Cf. Staal 2006: 90.
[20] However, note that if we follow the hypothesis that human languages first developed as gestural languages (much as the present-time sign-languages of the hearing-impaired) then spoken languages too can be viewed as 'artificial' in that they would have developed as a cultural product similar to writing.

At the same time, writing is obviously 'natural' in that human beings *can* do it. For the purposes of writing, humans clearly rely on cognitive possibilities that are 'naturally' present in their cognitive make-up from the start, as spelled out in detail by Dehaene (2009) and discussed in Chapter 5. But then, if writing is a 'natural' technological development, in what sense is it different from the technological/scientific development of, for example, mathematics or biology, and of the specific languages accompanying such developments? As Wang (1955: 236) aptly put it, '[s]o far as the development of human scientific activities is concerned, the creation of the language of the classical mechanics or of the axiomatic set theory was rather natural.' This is the case both from a historical and from a cognitive point of view; from both perspectives, we observe a gradual development in several steps towards these technological and scientific inventions. Moreover, the development itself, while of course being a 'cultural' development, is very 'natural' within the broader picture of human (social) cognition.

These considerations seem to lead to the conclusion that the distinction between so-called natural and so-called artificial languages on the basis of the criteria just discussed is at best one of degrees. This holds also of formal languages and formalisms; in particular, it is hard to see on what grounds one can attribute 'naturalness' to other forms of writing but refrain from attributing a similar form of 'naturalness' to formal languages.

More generally, the very natural v. artificial dichotomy is problematic in many senses. As argued by Bensaude-Vincent and Newman (2007), these terms seem to have ever-changing meanings, which are by and large relative to situations and backgrounds. Moreover, just as in the specific case of languages, these authors suggest that 'instead of opting for an absolute distinction of quality between the artificial and the natural, one should accept only a gradual distinction of degree' (Bensaude-Vincent and Newman 2007: 2). Still, Bensaude-Vincent and Newman believe that, given the pervasiveness of these concepts, we should not simply dismiss them as utterly inadequate; we should instead try to trace and systematize these different uses.

My take on the matter is perhaps more radical; given how unhelpful and misleading the natural v. artificial dichotomy has been in several specific debates, I believe that it is best forgotten. It remains of course a subject for historical inquiry, but as a philosophical/conceptual tool it is rather useless. Here are some of the debates in which variations of this dichotomy have done considerable harm:

(1) The nature v. nurture debate in biology. After much spinning around, it is now widely agreed that nature and nurture are inescapably intertwined

in both the ontogenetic and phylogenetic development of individuals and species. 'Internal', genetic features of an individual (nature) interact constantly with 'external' factors (nurture) in all biological processes; we have both processes of 'nature via nurture' (Ridley 2003) and conversely of 'nurture via nature'.

(2) The innate v. acquired debate in language evolution. After decades of debate on the existence or non-existence of a purported innate 'language faculty', many theorists now seem to think that such discussions benefit from discarding the dichotomy (e.g., Stenning and van Lambalgen 2008: chap. 6). No one denies that the 'internal' constitution of human beings is decisive for the development of the specific signalling systems that are human spoken languages. Indeed, even with extensive training, no other animal has ever come anywhere near becoming 'fluent' in one of the human spoken languages. But no one denies that external factors, in particular richness of stimulus and social interactions, are necessary conditions for the development of the ability to speak in a human being. As is often the case, the devil is in the details of how exactly these two kinds of factors are combined to give rise to the phenomena in question, but it is widely accepted that a comprehensive account will have to include both.

The bottom line seems to be that anything that human beings do (speaking, writing, walking, flying aeroplanes) must be constrained by the 'natural' possibilities of our cognitive and physical make-up, and are thus in some sense 'natural'. At the same time, given that we are inherently social animals, we quickly tend to develop particular ways to deal with each other and with the environment that are not in any sense 'hardwired' in us from the start. Likewise, it is now known that some non-human animals also develop forms of 'culture', such as different hunting techniques, different signalling calls, etc. (Rendell and Whitehead 2001), so the claim that 'cultural' is a concept to be applied exclusively to human beings becomes hard to maintain.

Now, back to languages: a wide body of evidence suggests that developing a highly complex signalling vocal system is something that is 'natural' in human beings. We have, to date, never encountered human populations who did not engage in what we call 'speaking'. Whether there are structural 'universals' in the multitude of human spoken languages is debatable, but that we all speak some form of language is *not* debatable. (But then, it is also known that without adequate stimuli at an early age a human being becomes severely impaired in her ability to learn to speak at later stages.) So, perhaps we could say that spoken languages are 'natural'. However, to what extent different *portions* of a spoken language are 'natural' becomes again a contentious issue: words for numbers, for example, are a relatively late development in most languages, and almost entirely absent in a number of human languages – though present in the wide majority of them (De Cruz and De Smedt 2010). Are numerals 'natural' or 'artificial'? It seems rather fruitless to formulate the debate in these terms.

In any case, it is clear by now that writing, more specifically, does not in any way qualify as 'natural' for a number of reasons: intensive training required for learning, rare historical appearances, not present in significant portions of the human population. So the traditional opposition between natural and artificial languages (the former also intended to cover 'everyday-life' written languages) is conceptually mistaken. Thus, we must look for alternative theoretical grounds in order to conceptualize the differences between formal languages and formalisms and other human languages.

2.1.4 *Language as practice v. language as object*

A typical speaker is a language *user* and does not routinely raise questions about this specific activity; it is just an organic part of her life which feels so natural and automatic that it does not require conscious reflection. True enough, in the course of formal education, we are at some point or another forced to look at language with different eyes. But, for the most part, language is essentially integrated in a broader range of practices, and in fact only 'makes sense' within this wider context.

In contrast, the language theorist has a different attitude towards her object of study. Of course, it could not be any different given that her job is precisely to reflect on these phenomena that other people engage in unreflectively. However, there is something slightly odd about a large chunk of this body of research: in the wide majority of cases, language is taken to be some sort of independent free-standing object, having autonomous existence vis-à-vis the human practices in which it is in fact firmly embedded. So, while it is to be expected that the theorist will go through a process of 'objectification' of her object of study,[21] the extent to which language studies are characterized by the view of language as an object, detached from the larger context of human practices and actions, is quite remarkable. Linell (2005: sect. 1.1) describes what he considers to be 'two ways of looking at language'. In his terms, these two ways are the following:

---

[21] Interestingly, this process of objectification typically involves focusing on written language at the expense of spoken language, as written language lends itself much more easily to the role of an independent, free-standing entity.

- Languages as structured sets of forms, used to represent things in the world.
- Languages as meaningful actions and cultural practices, interventions in the world.

According to Linell, the first conception of language has been predominant in the history of language studies: from Aristotle and Priscian to Saussure and Chomsky, most linguistic theorizing has been based on a conception of language as an independent entity, an inventory of forms detached from its actual embedding in people's lives and practices. Besides the syntactic characterization of forms generated by a certain number of rules, the semantic characterization of language according to this conception relies on the idea of language representing things and states-of-affairs in the world. Indeed, this approach tends to focus on a representational conception of meaning as described in 2.1.1 above. Crucially, the relation of representation does not require any human action in order to establish itself; a quintessential example of this approach is Wittgenstein's *Tractatus* (at least under one possible interpretation of the text, which notoriously allows for conflicting interpretations), where propositions are presented as pictures of states of affairs regardless of human practices or intentions. I will refer to this approach as the *language-as-object* approach.

In contrast, a minority of language theorists, while still to some extent inevitably 'objectifying' their object of study, recognize that it only makes sense to study language against the background of a wider context of practices and actions. Here again Wittgenstein, but this time the Wittgenstein of the *Philosophical Investigations*, is a prototypical example, but other theorists, also outside philosophy, have taken a similar stance – Tomasello's (2003) 'usage-based theory of language acquisition', for instance, can be viewed as falling under this category. It is tempting to regard the difference between these two attitudes as amounting to the role that pragmatics should play in language studies. But the divide cuts deeper than this, and concerns more generally the extent to which language use is viewed as embedded in human practices broadly construed. I will refer to this approach as the *language-as-practice* approach.

Let me comment briefly on the connections between the language-as-object v. language-as-practice distinction and the representational meaning v. usage-based meaning distinction introduced in 2.1.1. It might seem that the two distinctions are equivalent, but in fact they are not; the language-as-object v. language-as-practice distinction pertains to the *meta-level* of the *theorist's attitude* vis-à-vis her object of study, namely language, while the representational meaning v. usage-based meaning distinction pertains to the phenomenon-level of how to understand what meaning is and how it comes about in the first place. Of course, if the theorist has a conception of meaning that is essentially usage-based, she is much more likely to adopt the language-as-practice perspective in her analyses. But a theorist who endorses a representational notion of meaning may still adopt a language-as-practice perspective if she thinks that the quintessential use of language is that of representing portions of reality (and may go on to study how this is articulated in actual practices). In the other direction, while it is true that a representational notion of meaning is conducive to the adoption of a language-as-object theoretical attitude, it may be possible to reconcile a language-as-object attitude with a usage-based conception of meaning (as I think some recent work in formal pragmatics seems to be doing). So, while there is a conceptual proximity between the two distinctions, they are not equivalent.

Why is the language-as-object v. language-as-practice distinction important for the present purposes? Well, the basic point is that here we are interested in formal languages not only as (mathematical) *objects* as such, but rather in the broader picture of how formal languages are *used* and the impact they have on practices. While the tendency to objectify language and to adopt the first stance described above is predominant in language studies in general, it is exacerbated when it comes to thinking about formal languages. This is essentially due to the fact that formal languages are indeed typically construed as *mathematical objects*, and this is one of the reasons why they fulfil one of the functions they are summoned to perform in logical investigations, namely to allow for meta-theoretical investigations (cf. Chapter 3). It is then all too tempting to view them *exclusively* as mathematical objects and thus to disregard their function(s) and their cognitive impact on the practices of theorists (logicians or otherwise).

In short, in the present study, I adopt the language-as-practice stance to reflect on formal languages in particular. Formal languages are more than mere mathematical objects; they are also tools and means that genuinely affect the practices of logicians and other theorists who make extensive use of formalisms.

2.2 WHAT ARE FORMAL LANGUAGES THEN?

On the basis of the conceptual framework laid down so far, in this section I present the gist of what is probably an idiosyncratic but (to my mind) compelling conceptualization of the notion of formal languages. In the remainder of the book I put this conceptualization to use and discuss it in more detail.

2.2.1 *Formal languages as* languages

So let us start with a somewhat obvious and yet important question: in what sense are formal languages really *languages*? Is this a misnomer? In section 2.1.1, I discussed what I take to be prototypical features of the concept of a language, positing that human spoken languages provide the paradigmatic cases. However, the characterization was not intended as a set of necessary and sufficient conditions for something to count as a language, so the fact that some phenomena not displaying one or more of these features can still count as languages is perfectly compatible with this account. Let us now see how formal languages fare with respect to the four main features discussed.

Speech. A first and important observation is that formal languages typically do *not* have natural spoken counterparts. If they are expressed in oral contexts at all, it is only with some twisting and turning, as they are from the start essentially written languages.[22] The question is whether this feature is sufficient for them to be considered *non*-languages.[23] An argument for the view that formal languages are not languages properly speaking would be that 'ordinary' written languages are languages only insofar as they are the visual counterparts of spoken languages. But this is of course the phonographic/phonoptic conception of written language that has been rejected above. Now, if formal languages are written systems on a par with any other written system, then the issue seems to be whether written systems in general should be considered to be languages. This is, to some extent, a merely terminological issue, but given that many different written systems seem to fare well with respect to the other core features of the concept of a language, it would seem premature to dismiss them as non-languages at this point just for this reason.

Notice that the observation that formal languages are written languages does not commit us to the exclusion of some diagrammatic languages that have been proposed for the study of logic from the class of 'formal languages'. They range from Peirce's Existential Graphs to Barwise and Etchemendy's Hyperproofs – Shin and Lemon (2001) provide an overview.

---

[22] 'so-called "formal languages" ... construct graphical systems *sui generis* which are all, at best, verbalized retroactively and verbalized only in a limited, fragmentary form' (Krämer 2003: 522).

[23] Krämer (2003: 522 n. 1) notices that only a few linguists have included mathematical–logical notations as forms of language.

Not every diagrammatic system will be a formal language, but those systems that display the basic mathematical properties of formal languages, in particular explicit and exhaustive rules of formation, can be viewed as fully fledged formal languages. Notice, however, that the differences between diagrammatic and discursive formal languages do seem to have cognitive impact, in particular in the process of learning logic (Stenning 2002). (This aspect will be discussed in detail in Chapter 5.) At any rate, if Krämer (2003) is right, iconicity is a crucial feature of *any* writing system, including the allegedly discursive ones, so the difference might again be one of degree rather than a sharp one.

Semantics. This is perhaps the aspect with respect to which formal languages seem least entitled to the honorific 'language'. Formal languages are typically (though not always) *uninterpreted* languages, i.e., languages whose signs (for the most part) do not bear semantic relationships with specific objects, phenomena, or concepts by themselves (see Chapter 6). An act of 'interpreting' the language (typically onto mathematical structures) is required for it to become 'meaningful'. To borrow Krämer's terminology again, formal languages can be viewed as products of a process of *de-semantification.*[24] They do not appear to have the basic semiotic features that we usually associate with languages, and they certainly do not display the kind of 'representational meaningfulness' referred to above. Does that mean that, at least with respect to the semantic aspect, formal languages are not languages, properly speaking? Well, I submit that there is a sense in which formal languages are in fact 'meaningful', but essentially in the sense of 'usage-based' meaningfulness. I will argue for this point in more detail shortly, when discussing the functionality of formal languages below.

Syntax. By contrast, if syntax is indeed a core feature of any language, then formal languages satisfy this criterion with flair. In effect, while actual spoken languages have a somewhat loose relationship to the systems of grammar that presumably underpin each of them, when it comes to formal languages, observance of the rules of formation is an all-or-nothing affair: only the expressions that follow strictly the rules of formation of a given formal language are to count as well-formed formulae of the language. Again, if syntax is such a crucial feature in all languages (which is a debatable point, as suggested above), then formal languages take this feature to the

---

[24] Notice that formal languages are in first instance typically developed against the background of a particular application or domain; to my knowledge there has never been a formal language introduced as a purely arbitrary collection of symbols and rules of formation. So, even if at a later stage formal languages can be treated as mere 'blueprints', in first instance they are always connected to some specific ideas or application.

next level; one might even be tempted to conclude that formal languages are 'superior' languages, at least from this point of view. This would be quite absurd, of course, which only goes to show that the centrality of grammar and syntax for spoken languages may be overestimated, as argued by Tomasello (2003) and others.

Communicative function. We have seen that plausible (but by no means indisputable) accounts of the functions of human (spoken) languages stress the communicative role they play in human interactions. We are social animals, and a complex system of signalling comes in handy in terms of coordinating the actions of the members of a group.[25] Now, to what extent do formal languages fit the picture of human languages as means for communication?

To some extent, they do. Just as with mathematical notation generally speaking, formal languages do serve the purpose of communication within the scientific community. Also, we cannot deny that the fairly standardized use of at least some of the most frequent logical signs simplifies contact among peers, enhancing precision and objectivity. Whether formal languages fulfil a *communicative/expressive* function better than other languages in specific contexts is a matter to be discussed in more detail in Chapter 3; but, for now, it is sufficient to notice that they do play a communicative role, at least to some extent.

However, this seems to me to be only the tip of the iceberg, so to speak. The crucial insight that this whole investigation is based on is the idea that formal languages play a functional role that goes much beyond communication: they trigger certain *cognitive* processes that would typically not be triggered in their absence. This hypothesis will be systematically put under scrutiny in Part II, but for now let me present it in a bit more detail in terms of the concept of *operative writing* as introduced by Krämer. Here is how she presents the concept:

The advantage of conceiving writing as non-phonetic reveals a whole new realm of written phenomena, which will be called 'operative writing' in contradistinction to phonetic writing. Calculus is the incarnation of operative writing. This term can be understood as a system of graphical signs comprised of a finite repertoire of discrete elements that serves a dual function. On the one hand, it is a medium for representing a realm of cognitive phenomena. On the other hand, calculus provides a tool for operating hands-on with these phenomena in order to solve problems or to prove theories pertaining to this cognitive realm. Thus, we can isolate a third

---

[25] But notice that there are many animals with highly complex social structures which do not rely on this particular kind of signalling system.

type of blindness inherent to the phonographical conception of writing: the omission of operative systems of notation from written phenomena. (Krämer 2003: 522)

Operative writing differs in many respects from other forms of writing. Crucially, operative writing is not primarily about *communicating*: it is not meant to express something to *another* agent but rather to bring about within-agent specific cognitive processes. Speech can have such a function too (e.g., counting out loud, internal speech), but there is something about the concreteness of writing that seems to be particularly conducive to triggering certain cognitive processes. Notice, however, that contrary to what Krämer seems to be suggesting, the notion of 'operative writing' is best understood not as covering a specific *class* of writing systems (formal languages but in fact formalisms more generally), but rather as a specific *function* that different writing systems may have. So-called 'phonetic' writing can also have an operative function in the sense of helping one organize one's thoughts, often not in fully discursive form but with abbreviations and diagrammatical representations. A 'to-do' list or a shopping list, for example, arguably constitutes instances of operative writing.[26] So, while it is natural to think of operative writing as epitomized by languages such as programming languages and the formal languages of logicians, it would seem incorrect to disregard the seemingly operative uses that are made also of other, more prosaic forms of writing.[27] (We'll come back to this issue in Chapter 6 and explore the issue of what distinguishes Krämer's notion of operative writing from Menary's notion of 'writing as thinking'.)

Moreover, operative writing can of course be used by a given agent in order to trigger similar cognitive processes in another agent (e.g., the instructor who writes down logical symbols during a logic class), but its primary use seems to be in mono-agent situations. In other words, the claim here is that the primary function of formal languages insofar as they are a form of operative writing is not a communicative one. Whether this disqualifies them as languages properly speaking will depend on whether the concept of a language is necessarily tied to a communicative function.

For the most part, formal languages and formalisms are not meaningful in a traditional, representational sense, i.e., in the sense of bearing special

---

[26] Think of Otto's notebook in Clark and Chalmers's (1998) famous thought experiment to argue in favour of the concept of an extended mind (EM) (the EM hypothesis is discussed below, in Chapter 5).

[27] See Menary (2007b) for a 'hands-on' conception of writing in general (also discussed in section 5.2.1, below).

semiotic relations with portions of the world. Moreover, if they are not geared towards communication either, it is unclear to what extent they may be said to be meaningful in a 'usage-based' sense. As described above, a usage-based conception of meaning is based on the effects that specific uses of portions of language bring about primarily with respect to other agents and in the world; now, formal languages as a form of operative writing do not seem to have either one of these two effects in any obvious way. If, however, one is prepared to extend the notion of usage-based meaning so as to cover the *within-agent* cognitive effects of using a portion of language, then perhaps a case can be made for the claim that formal languages do have a usage-based kind of meaningfulness, at least if broadly construed (i.e., insofar as they are constitutive elements of meaningful human practices).

So let us now take stock: in terms of the four characteristics of a language just presented, to what extent can we say that formal languages are *languages*? They clearly have no natural speech counterparts; their semantic import is not representational, and only on a very broad conception of semantics can they be said to be 'meaningful'; in contrast, in terms of syntax they actually seem to 'fare better' than any other non-formal language; finally, their function is not primarily a communicative one, as they seem to be particularly suitable to trigger within-agent cognitive processes.

Do formal languages qualify as languages on the basis of these features? Again, it will all depend on whether one is prepared to stretch the concept of 'language' beyond its familiar borders. This is not a particularly interesting question, though, and defending the point that formal languages should be considered as fully fledged languages is not part of the agenda here. Instead, what *is* important is to bear in mind these basic properties of formal languages as they emerge from the systematic comparison with more 'familiar' kinds of languages (human spoken languages in particular).

2.2.2 *Formal languages as* formal

While nothing essential seems to hinge on formal languages being considered as languages or not (it is mainly a matter of terminology, a verbal dispute), the same does not hold of the attribution of a 'formal' character to these languages. It is crucial that we understand in which sense(s) formal languages are *formal*, and what exactly this tells us about logic and its 'formality'.

Let me start by noting that a compelling case can be made for the idea that to be formal in general is essentially a matter of degrees; this can be argued for from a variety of perspectives. For example, if the concept of the formal is associated with processes of abstraction (both the processes themselves and their products), i.e., of separation of matter from form, then presumably such processes can be carried out to different degrees of completion. Generally, many of the different variations of the formal I have discussed in Dutilh Novaes 2011b allow for degrees of being formal, in such a way that the formal v. non-formal divide would not be a sharp, clear-cut demarcation.

Yet, it seems that the concept of a formal language is *not* compatible with the idea of a distinction of degrees. True enough, a language can be *regimented* to a greater or lesser extent, but the concept of a fully fledged formal language is a mathematically well-defined concept with precise borders. A structure is a formal language iff it is defined by a finite collection of symbols and rules of formation which determine exactly what is to count as a permissible combination of symbols, i.e., a well-formed formula of the language. (In other words, the definition of a well-formed formula is recursive with respect to the formation rules.) Thus defined, a formal language is a *mathematical object*, and its properties can be (mathematically) investigated. In practice, however, much of the work done in mathematics and in logic (both 'pure' and applied logic) does not really make use of fully fledged formal languages, using instead collections of symbolisms and notations that could perhaps be described as 'semi-formal'. The main difference would be that in the latter case no explicit formulation of formation rules is given, or in any case not exhaustively. (In the present investigation, we are interested both in fully fledged formal and in semi-formal languages.)

Moreover, once the class of well-formed formulas of a language is defined, rules of *transformation* may be specified which determine the legitimate steps from one formula to another.[28] When such rules of transformation and axioms are formulated in the same language, this yields a *formal system* or *formal theory* having the language in question at its basis. (Obviously, the relation between formal languages and formal systems/theories is one to many.) Now, when only the rules of transformation are specified, the result is often referred to as a 'logic', which is then amenable to be used as the underlying logic of different formal theories (i.e., coupled

---

[28] In *The Logical Syntax of Language* (to be discussed in Chapter 3), Carnap (1934) views the totality of rules of formation and of transformation as a language, as a calculus, thus including the latter in the concept of a language. But, in general, the rules of formation alone are thought to characterize a formal language.

with different sets of axioms).[29] Although we here focus chiefly on formal *languages*, the truly interesting cases are those when formal languages are actually put to use underlying formal systems and logics.

Let us now consider what is *formal* about formal languages on the basis of the two variations of the formal discussed in Chapter 1.

Formal as de-semantification. This is clearly one of the main senses in which formal languages are thought to be formal. Even though there exist of course formal languages which are meaningful and thus not uninterpreted languages (e.g., Martin-Löf's constructive type theory), formal languages as they are currently used are predominantly uninterpreted languages. Notice also that the idea of manipulating symbols as blueprints with no meaning at all by themselves is not sufficient to constitute a formal language (or a formal system/theory); explicit and exhaustive rules of how to manipulate the symbols are also required. In this sense, the 'formal' conception of arithmetic as championed by Thomae and others in the nineteenth century is already a (proto-)variation of formal as de-semantification, but it is not based on an underlying fully fledged formal language.

Again, the concept of 'de-semantification' is borrowed from Krämer (2003). She presents the idea of 'operative writing' as a particularly vivid illustration of the process of de-semantification: 'signs can be manipulated without interpretation'. As we have seen above, according to her, calculus is an ideal example of the process of de-semantification:

The rules of calculus apply exclusively to the syntactic shape of written signs, not to their meaning: thus one can calculate with the sign 'o' long before it has been decided if its object of reference, the zero, is a number, in other words, before an interpretation for the numeral 'o' . . . has been found that is mathematically consistent. (Krämer 2003: 532)

It might seem awkward to use the term 'de-semantification' with respect to formal languages insofar as their components are presented as 'meaningless signs' from the start – that is, as mathematical rather than as semiotic objects. But I take it that what 'de-semantification' refers to in this case is the more general *historical* development from our usual tendency to operate and reason with meaningful languages towards the technique of reasoning with (manipulating) meaningless signs.

---

[29] Notice, though, that there is a certain fluidity in the concepts of axioms and rules of transformation within a formal system. A logic can be formulated in an axiomatic form with a very minimal set of rules of transformation (typically, only *modus ponens*), or these very axioms can be formulated as rules of transformation.

Besides offering a terminological reason why formal languages are indeed formal, the de-semantification aspect of reasoning with formal languages will be of absolutely crucial significance for the explanation of the cognitive impact of using them. I urge the reader to keep this in mind as we move on; the concept of de-semantification will be discussed in detail in Chapter 6.

Formal as computable. While the formal as de-semantification is crucial for the cognitive import of uses of formal languages, but is neither a necessary nor a sufficient condition for a language to count as formal, the formal as computable does constitute at least a necessary condition for something to count as a formal language. Moreover, it is also a decisive aspect of the cognitive resources afforded by formal languages and formal systems/theories more generally. Indeed, it is precisely the combination of de-semantification with the application of (external) rules of transformation that allows formal languages and formalisms to be such powerful cognitive tools. What it means for a language to be formal in the sense of computable will be given an interpretation from the point of view of extended cognition in Chapter 5. Essentially, I will argue that the explicit formulation of rules of transformation results in an *externalization* of reasoning processes, which moreover rely heavily on sensorimotor processing by the agent (literally moving bits and pieces of the notation around).

As already mentioned, the general idea of the formal as computable operates on two different levels: the level of how the lexicon of a given language is combined to generate well-formed formulae on the basis of the syntax of the language (its formation rules), and the level of the rules of transformation of the logic (or formal system) that are built upon the formal language in question. In the first case, the procedure is computable in the strict sense that it can always be decided on a finite number of steps (i.e., assuming no infinite formulae) whether a given formula is well formed or not. In the second case, however, the procedure is not always decidable in that there may not be an effective, recursive procedure to determine whether a given formula can be attained by means of a finite number of (inferential) steps from a given set of starting-point formulae (premises). Nevertheless, even in undecidable systems, a significant number of con-clusions can be drawn by strict applications of the rules of transformation/ inference of the system; strictly speaking, the latter cases do not qualify as 'computable' in the sense of the availability of a general decision procedure, but they do display a similar pattern for the application of rules.

In any case, a formal language strictly speaking is generally character-ized by the fact that its well-formed formulae are recursively enumerable; when this condition fails, we can at best speak of semi-formal languages.

Thus, I submit that it is chiefly in these two senses of the formal, as de-semantification and as computable, that formal languages are formal; and precisely these two features seem to be the main causes of their strength as cognitive tools.

2.2.3 *Formal languages as a technology*

We are now better equipped to go beyond the strictly mathematical definition of a formal language and to view this phenomenon as embedded in a larger context of human practices. Recall that in section 2.1.4, I contrasted the language-as-object approach to the language-as-practice approach in language studies, and stated the goal of adopting a language-as-practice approach with respect to formal languages. In Chapter 3, I discuss in more detail the technical developments that led to the conception of formal languages as 'object-languages'; it is a perfectly legitimate conception, and one which afforded many crucial technical results. So the aim here is not to supplant or discard the language-as-object approach but rather to offer an alternative, complementary viewpoint on formal languages.

The basic idea is that formal languages can be fruitfully conceived of as a *technology*. Of course, this is not very informative unless we can provide a more precise meaning to the rather vague term 'technology'. As a first approximation, a technology can be described as a specific method, material, or device used to solve *practical problems*. Formal languages as such are not a method by themselves, but they are devices that allow for the implementation of certain methods.

An important aspect of technologies in general is that they are typically developed with specific applications in view, i.e., to fulfil certain needs that the hitherto available technologies fail to address; but they often turn out to offer possibilities that had not been originally foreseen, and which go beyond the specific practical problems they were created to address. We could refer to this phenomenon as the 'surprise factor' in any technological development. What this means is that, once a given technology is developed to tackle a specific familiar issue, it often opens up a whole new range of possibilities and applications which would otherwise not even be conceivable prior to its development. New technologies may literally create new worlds, besides bringing answers to existing problems.[30]

---

[30] Needless to say, the surprise factor in technological development can also lead to disastrous consequences.

We may draw an analogy between surprising applications of a given technological development and the concept of *exaptation* from evolutionary biology, as introduced by Gould and Vrba (1982). Indeed, in the evolution of living beings, it frequently happens that a trait can evolve because it served a particular function at a given time, but subsequently it may come to serve another function, and this phenomenon is described as exaptation. Now, when it comes to technologies, something like exaptation seems to occur quite often: a technology is developed to fulfil a given function or functions, but at a later stage it is co-opted for applications that were not originally foreseen by its developers.[31]

Perhaps the best example of a radically life-changing technology is writing.[32] (Speech could probably also count as a technological innovation, but its emergence is so remote that it remains a controversial and obscure issue.) As we have seen, writing developed over thousands of years in Mesopotamia, but for the longer part of this development it consisted in very rudimentary forms of proto-writing. Originally, proto-writing was developed as a simple technology for accounting, i.e., for keeping track of production surplus and goods in general; naturally, such needs only arose against the background of a society which had abandoned a hunting–gathering economy in favour of agriculture and herding.[33] So, what started as rudimentary techniques to keep track of goods slowly but surely developed into a much more powerful technology, one which profoundly modified the way human beings (in literate societies) live their lives.

Of course, formal languages *are* written languages, so the history of their development is a specific chapter in the general history of the development of writing. An important aspect thereof is how the development of mathematical notation in particular went hand in hand with the search for notations that would facilitate processes of *calculation*; this is especially conspicuous in the development of numerical notation – see De Cruz and De Smedt 2010. Interestingly, with respect to numerical notation, expressivity and ease of calculation often do *not* go hand in hand. The classical example is the contrast between Roman numerals and Hindu–Arabic numerals; the former are in a sense more intuitive and 'iconic', but they constitute a cumbersome tool for calculation, while the opposite is true of

---

[31] Obviously, the fundamental difference is that, in the case of technologies, there is a specific agent behind their development, with a specific design in mind, which is of course not the case of evolution as an essentially random process.

[32] See Krämer 2003 on writing, and operative writing in particular, as a *cultural technique*.

[33] As already mentioned, writing did arise independently at a few other places and times, but the history of such developments is not significantly different.

the latter.[34] Indeed, there seems to be an inherent tension between desiderata of expressivity and desiderata of effective calculation, a tension which
will occupy a central position in subsequent discussions (especially in
Chapters 3 and 7).

In the next chapter, I present an overview of the historical development
of this technology, emphasizing at each point the practical needs and
avowed goals that motivated the search for new tools to address them,
thus developing the technology of formal languages. My general claim will
be that the ultimate actual impact of this technology on the practices of
logicians seems to go well beyond these avowed goals, but that we still hold
on to them more than we should when reflecting on the methodology of
formal methods, thus failing to attend to the cognitive impact of using
formal languages. Indeed, I will attempt to outline what exactly is the
'surprise factor' of formal languages as a technology that makes them such
powerful cognitive tools.

2.2.4 *The cognitive status of formal languages*

In effect, a leitmotiv in the recent history of this technology is its presumed
expressive advantages. Formal languages are typically presented as more
precise and conspicuous, and thus as better tools for the *expression* of
scientific theories. Again, it is worth noticing that concerns of expressivity
pertain predominantly to multi-agent, dialogical situations, that is, situations in which one agent is communicating with different agents.[35] Now, as
already mentioned, it is not so much that formal languages do *not* play this
role, but rather that this is a secondary aspect of their impact. Indeed, the
real power of formal languages – what makes them truly indispensable –
seems to lie in the cognitive impact they have as a paper-and-pencil device, a
tool for operating hands-on with cognitive phenomena. As such, the
technology is typically used in mono-agent contexts – even though logicians

---

[34] See Krämer 2003: 531. But, recently, an efficient algorithm to calculate with Roman numerals has
been developed (Schlimm and Neth 2008), which at any rate shows that the claim that it is *impossible*
to calculate efficiently with Roman numerals is incorrect. What remains true is that until now no such
procedure had been found, and that the move from Roman numerals towards Hindu–Arabic
numerals was (among other reasons) historically motivated by the search for improved calculating
methods.

[35] This is not entirely true of foundational projects such as Frege's logicism. In such cases, the search for
epistemically firm grounds is the main motivation for the development of his ideography; it was not
intended as a language to write proofs to be presented to others. It does end up having a strong
operative dimension though, perhaps even unbeknownst to Frege himself (cf. Macbeth 2011).

and mathematicians do often 'brainstorm' in larger groups, usually while frantically writing formulae on whiteboards or similar surfaces.

The basic idea, which will be argued for on the basis of empirical data in Part II, is that formal languages are a technology that allows us to reason in ways that are fundamentally different from how we spontaneously reason in more mundane circumstances. We tend to rely on what has been described as a 'fundamental computational bias of human cognition', that is

[t]he tendency to automatically bring prior knowledge to bear when solving problems . . . this tendency toward contextualizing problems with prior knowledge is so ubiquitous that it cannot easily be turned off – hence its characterization here as a fundamental computational bias (one that pervades virtually all thinking, whether we like it or not). (Stanovich 2003: 292–3)

The hypothesis to be developed here is that formal languages allow us momentarily to 'turn off' this computational bias and to run different software, as it were. They are external devices (in the sense of extended cognition – see Chapter 5) giving rise to different patterns of reasoning, which, in turn, especially in scientific contexts, may enhance the discovery of new facts. This happens precisely because in such contexts it may be advantageous to compensate for some of our spontaneous reasoning tendencies, in particular the tendency of bringing prior knowledge to bear during the reasoning process.

Now, most recent proponents of the use of formal languages (and methods) in the history of their development seem to have overlooked this particular effect. One needs to go back to the seventeenth century, with Leibniz in particular, to find statements of the idea of formalisms serving as tools for discovery (see section 3.1.1).[36] So the general claim here is that, since Frege at least, formal languages as a technology developed essentially against the background of avowed needs related to expressivity, but that they in fact have a much deeper, powerful cognitive effect in how we reason when resorting to them. This particular feature of formal languages has never been explored in detail, and the present investigation aims at addressing this important lacuna.

All this will be spelled out in detail in Part II, but for now it seemed convenient to advance the gist of the main hypotheses underlying this investigation, given that they are deeply connected with the reconceptualization of formal languages presented in this chapter.

---

[36] There is a sense in which the Carnap of *The Logical Syntax of Language* (1934) is also endorsing a similar idea, but his motivations are not related to cognitive concerns.

2.3 CONCLUSION

Thus, for the continuation of this investigation, the most important features of a formal language according to the conceptualization just presented are: they are *written* languages; their semantic dimension is not representational but rather usage-based *within-agent*; their function is predominantly operative rather than expressive; they are *formal* in that they are the product of a process of de-semantification and in that they display key computational properties (which in turn hinge crucially on their status as *external* cognitive artefacts).

For this reconceptualization of formal languages to emerge, a series of premises and assumptions had to be put under scrutiny, such as: a re-evaluation of the status of writing vis-à-vis speech; the adoption of a language-as-practice approach; a broadening of the notion of meaning so as to include what I described as 'interventional meaning'; a dismissal of the natural v. artificial dichotomy with respect to languages (and in fact more generally).

I am well aware of the somewhat idiosyncratic character of this conception of formal languages, and also of the fact that they do not seem to reflect much of the current practices of logicians (resonating more with an old-fashioned Leibnizian conception of formalisms); currently, logicians do not seem to use formal languages to reason *with* but rather predominantly to reason *about* (see Chapters 3 and 7). But I believe that the reader's willingness to accept it as a working hypothesis for now will be rewarded in the next chapters, where a more general methodological reflection on uses of formal methods will be presented on the basis of the 'formal languages as cognitive technology' hypothesis.

CHAPTER 3

The history, purposes, and limitations of formal languages

As the title indicates, this chapter focuses on the history of formal languages, offering a condensed account of their historical emergence on the basis of the hypothesis of formal languages as a cognitive technology (Chapter 2). As will become apparent, we will have to go beyond the scope of logic as a discipline and examine the historical development of *mathematical* notations. In addition, this chapter also offers a (primarily historical) discussion of the purposes that are traditionally attributed to formal languages; I will argue that the typical rationales given for the use of formal languages follow three basic lines: expressivity, iconicity, and calculation. Finally, I discuss some limitations and misuses of this technology; as with any tool, a formal language can also *malfunction*, either because it is not suitable for a particular application, or simply due to the inherent limitations of formalisms (I discuss incompleteness and non-categoricity in particular). All in all, this is essentially a historical chapter, but the points discussed will be crucial for the systematic and empirically informed analysis of subsequent chapters.

3.1 THE HISTORICAL DEVELOPMENT OF A TECHNOLOGY

In this section, I offer a brief survey of the historical developments leading to the emergence of formal languages. As suggested in Chapter 2, the key hypothesis is that formal languages are best viewed as a *cognitive technology*, developed in order to facilitate a range of cognitive processes; but, naturally, this facilitating effect can be understood in a variety of ways, as will become clear.

In effect, what follows is a modest attempt at what Netz (1999) has described as 'cognitive history': a historical analysis which takes into account the cognitive background, motivations, and implications of the developments in question. More specifically, the underlying idea is that the historical development of formal languages is best understood from the point of view of the *extended cognition* framework (developed in more detail in Chapter 5). In fact, it would seem that the whole history of notations in mathematics could (should?) be written from the point of view of the concept of extended cognition, but this more ambitious goal falls out of the scope of the present investigation. For reasons of space, the survey here is rather brief, but it emphasizes the role of the development of algorithmic and algebraic techniques for calculation in the Arabic world (against the background of progress in Indian mathematics), which were brought to Europe by the abbaco schools. Without this link, it is impossible to understand the progress in mathematical notation in the sixteenth and seventeenth centuries, initiated by Viète and completed by Descartes.

At any rate, a passage by cognitive scientists Landy and Goldstone illustrates particularly well the point of view to be adopted here:

Over the course of history, mathematical formalisms have evolved so that they are cognitively helpful devices, and this evolution has entailed making apparently superficial, but practically crucial, form changes. For example, the convention introduced by Descartes (Cajori 1927) in which letters near the beginning of the alphabet are used to denote constants and those near the end to denote variables, frees us from the burden of remembering which are which and allows us to use our memory for other aspects of a mathematical situation. (Landy and Goldstone 2007a: 2039)

Some of the questions to be kept in mind are: What were the immediate goals practitioners had in mind when developing a given new technology/technique? What were the unforeseen applications of these novel developments, if any?

3.1.1 Early developments[1]

The tradition in logic that is of primary concern to us here begins in ancient Greece, more specifically with Aristotle. However, to understand the relevant developments in notation ranging over roughly 2,500 years, we must also consider developments which took place outside Europe, in particular in India, China, and the Arabic world. As described well by Staal (2006), the history of formal languages (he uses the phrase 'artificial languages') spans through thousands of years and at least two continents. More generally, it must be examined against the background of the development of notations in *mathematics*, and in particular the search

---

[1] This section has greatly benefited from comments by Albrecht Heeffer.

68   *Part I*

for notations that could serve as calculation instruments.[2] Indeed, many
of the central developments in notation took place within mathematics,
and were only later (essentially in the seventeenth century) imported into
academic logic. We may say that, from its birth in ancient Greece up to
the seventeenth century, academic logic was basically practised with a
steady body of notational devices and techniques, in particular schematic
letters and regimentation.

Another strand which seems to have greatly contributed to the notational
revolution of the seventeenth century is the Lullist tradition, inspired by the
revolutionary ideas of Ramon Lull (1232–1316). Lull is himself a product of
the rich confluence of cultures in the medieval Mediterranean area, being
conversant with the Islamic, Jewish, and Christian traditions (Lohr 2010).
Medieval (academic) scholasticism remained by and large oblivious to Lull's
ideas on the symbolization and mechanization of knowledge, but it is now
well known that he exerted significant influence on authors such as
Nicholas of Cusa, Giordano Bruno, Gerolamo Cardano (each treated in
Lohr 1974), and even Leibniz. However, for reasons of space, we will here
mostly focus on the contributions of the development of algebraic notation,
but with the caveat that a complete account of the history of formal
languages would also have to take the Lullist tradition into account
(which will be discussed rather briefly).

Before discussing schematic letters and regimentation specifically, a first
observation is in order. Both for logic and for mathematics, a key transformation in ancient Greece was the transition from *oral* to *written* contexts.
As argued by Netz, the birth of the deductive method takes as its starting
point purely oral, dialogical situations, which then become regimented in
written forms.

Greek mathematics reflects the importance of persuasion. It reflects the role of
orality, in the use of formulae, in the structure of proofs ... But this orality is
regimented into a written form, where vocabulary is limited, presentations follow a
relatively rigid pattern ... It is at once oral and written. (Netz 1999: 297–8)

The observation applies, *mutatis mutandis*, to logic as well, which emerged
as a codification of dialogical practices in the Academy (Marion and
Castelnerac 2009). The transition from the oral to the written medium

[2] Staal (2007a: 578) attributes 'the origin of artificial languages to a fusion of two innate faculties, that of
language and that of counting or mathematics generally.' This claim, thus formulated, is quite
contentious (especially with respect to the concept of 'innate faculties'), but the general idea that
formal languages arise at the intersection of mathematical and linguistic practices is at the basis of the
present analysis, and Staal's piece provides further historical evidence for it.

is epitomized in Plato's dialogues, which registered and passed on to future generations Socrates' orally formulated teachings (despite Socrates' own misgivings vis-à-vis the written medium). The point is not that regimentation is only possible on the written medium, as might be thought; Staal (2006) has argued convincingly that in the Indian grammatical tradition one finds theorizing and regimentation of purely oral contexts. But, in the case of ancient Greek logic and mathematics, there is a clear connection between regimentation and the transition to the written medium.

What I mean here by regimentation is made particularly evident in a passage from the *Prior Analytics*, in a reworked version as presented by Hodges (2009: 592).[3] Consider the argument:

(6) God doesn't have times that need to be set aside for action. God does have right moments for action. Therefore some right moment for action is not a time that needs to be set aside for action.

Aristotle would find the premises and the conclusion, and then write out the syllogistic terms together with letters to represent them:

(7) A: thing that God has. B: time needing to be set aside for action. C: right moment for action.

This procedure would lead to the following regimentation of the original argument (Hodges 2009: 592):

(9) No (time needing to be set aside for action) is a (thing that God has). Some (right moment for action) is a (thing that God has). Therefore some (right moment for action) is not a (time needing to be set aside for action).

The formulation of a 'translation key' associating portions of the original argument to schematic letters is not a necessary component of regimentation: the direct move from (6) to (9) would already count as regimentation. But why does Aristotle go through the trouble of reformulating an argument in this way? Well, simply because the argument as originally formulated would not be amenable to analysis with the syllogistic machinery. But, once it is rewritten, the (valid) syllogistic mood 'No B is A. Some C is A. Therefore some C is not B' naturally emerges from the argument. Regimentation and the use of schematic letters yield *schemata*, and for most of its history (and still now, at least on certain levels) schemata were viewed as objects of logical analysis par excellence.

---

[3] Hodges: 'This is from Prior Analytics i.35f. Aristotle's text needs careful dissection, and for a smooth exposition I've permuted some of his material.'

Thus, this style of regimentation marks the very birth of logic with Aristotle's syllogistic.[4] As is well known, syllogistic is a system with relatively limited expressive power, as it only deals with four sentential structures (in its original formulation), also known as categorical sentences: 'All A is B', 'Some A is B', 'No A is B', and 'Some A is not B', where the schematic letters 'A' and 'B' can be replaced by terms of the appropriate kinds.[5] Thus, if syllogistic is to have wide-ranging applications, it must be possible to rewrite arguments so that the 'appropriate' sentential structures emerge.

While syllogistic has occupied a prominent position throughout the history of logic, it was not the only logical theory requiring accommodations/regimentations of this kind. In fact, it is a familiar story: a given logical system recognizes a small number of patterns in the (typically linguistic) phenomena that it is supposed to address, and those instances of the phenomena which do not fall straightforwardly into one of the recognized patterns must either be regimented or declared out of the scope of the system. At this early stage, regimentation is thus closely related to the very possibility of logical theorizing: the idea is to work with a small class of patterns for reasons of feasibility, but to aim for these patterns to 'cover' as many cases as possible, so that the theory can be widely and fruitfully applied.

Besides syllogistic, another example worth mentioning is the case of Latin medieval theories of supposition, which were theories intended to provide a systematic account of the semantic behaviour of terms and sentences.[6] Just to mention one such example, Latin medieval logicians adopted conventions related to word order so as to disambiguate some constructions: 'Every man loves a woman', for example, would be written 'A woman every man loves' if the intended reading was (in anachronistic terms) to assign wider scope to the existential quantifier. (In medieval terminology, in the first case 'a woman' would have merely confused supposition, while in the second case it would have determinate supposition.) More generally, Latin medieval logic in general is characterized by a high level of regimentation of the

---

[4] Interestingly, since the 1980s, the 'natural logic' movement has been promoting the study of logical reasoning in what is described as 'natural language', essentially relying on syllogistic systems (or variations thereof). It is ironic that the very birth of syllogistic is in fact marked by a move *away* from ordinary language; it is widely acknowledged that the categorical sentences of syllogistic were rather contrived even from the point of view of the Greek spoken at the time. Hodges (2009) very appropriately asks: What is *natural* about so-called 'natural logic'?

[5] Notice that Aristotle did not use the 'is' copula but rather the term 'belongs to', with an inversed order of subject–predicate. So 'No A is B' is originally formulated as 'B belongs to no A', etc.

[6] I have given some examples of such regimentation in Dutilh Novaes 2008.

language used. In fact, academic Latin was not anybody's 'native language' at that point; it had become first and foremost a tool for intellectual inquiry.

A regimented language is not yet what we would typically describe as a formal language, but it certainly constitutes a conceptual step towards the development of formal languages (although, as we will see, 'our' formal languages did not have their main historical roots in the logical practice of regimentation). By contrast, the systematic use of schematic letters is from the start something we easily recognize as a 'formal component' in a language, and from this point of view even the original categorical language of syllogistic can be viewed as a 'formal' (albeit extremely simple) language.

Historically, again there seem to be interesting connections between the use of letters in logic and in mathematics in ancient Greece. Netz (1999: chap. 1) explicitly relates the use of schematic letters by Aristotle in the two *Analytics* to the emerging practice of using letters to denote points in a diagram ('the lettered diagram', as he calls it). He hypothesizes[7] that Aristotle may have been inspired by the (then already) well-established mathematical practice of using letters in diagrams for the introduction of schematic letters in his logical work.[8] In effect, the generality and arbitrariness brought in by the use of letters in diagrams ('Let ABC be a triangle defined by the points A, B, and C') may have been Aristotle's inspiration to use letters to designate arbitrary individuals, thus obtaining generality. Indeed, the concept of a schematic letter (placeholder) must be sharply distinguished from the concept of a *variable*, as it emerged within mathematics: a mathematical variable stands for an unknown but determinate value whereas a schematic letter is a device of generality, indicating a range of possible instantiations.[9]

Be that as it may, the use of schematic letters is arguably one of the main reasons why syllogistic is a logical system even by modern standards. It deals with *schemata*, so that the principles and rules stated are valid for any (permissible) instantiation of the schemata with specific terms. This feature also allows for meta-properties of the system to be investigated, as is done in the *Prior Analytics* (e.g., that every valid syllogism can be reduced to a perfect syllogism, among others).

---

[7] On the basis of grammatical–linguistic considerations (Netz 1999: 48–9).

[8] It is interesting to notice that the status of Aristotle's logical works as *written* work is debatable; the hypothesis that his texts are in fact notes taken by students during classes is viewed as quite plausible (which would explain the 'dry' style, as opposed to the much more literary style of Plato's dialogues). Again, this would suggest that the writing medium is not a necessary condition either for regimentation or for the use of schematic letters.

[9] Moreover, in the modern usage of these concepts, schematic letters range over terms, and variables range over objects of the domain.

The use of schematic devices remained pervasive in the history of logic, in particular but not exclusively in connection with syllogistic. In fact, a different ancient Greek tradition in logic (just as remarkable, but having had significantly less historical influence) also made extensive use of schematic devices: the Stoic tradition. The Stoic logicians systematically used *numerals* to formulate patterns of valid reasoning, but, rather than standing for terms such as in Aristotelian syllogistic, the Stoic schematic devices stood for sentences/propositions.[10] Modus ponens, for example, was thus formulated: 'If the first, then the second; but the first; so the second' (modus tollens: 'If the first, then the second; but not the second; so not the first').

It is fair to say that, in terms of linguistic devices, regimentation and schematic letters were the main notational techniques used by logicians from Greek antiquity up to the seventeenth century in the so-called 'Western' tradition. Through the centuries, the use of schemata became increasingly associated with the notion of the *formal*, giving rise to what can be described as the schematic conception of the formal.[11] Now, to understand the notational technological explosion that took place in the seventeenth century, one must look elsewhere for its historical sources: outside Europe, and outside logic. So let us turn to these developments now.

It may not seem obvious at first sight, but the first step in the historical developments leading to the emergence of formal languages is arguably the appearance of place–value numerical systems. Historically, the first such system appeared with the Babylonians, but some prominent ancient mathematical traditions did not have place–value systems, including the Greeks and the Romans. The first fully fledged place–value numerical system emerged in Indian mathematics, starting in the first century. Indian mathematicians developed a numerical framework which was able to express extremely large numbers; they were also the first who conceptualized zero as a number (emerging from the notational device to indicate 'gaps', which is required for a place–value system).[12]

As is well known, the Indian place–value system made its way to Europe via Arabic mathematics. In particular, algorithms for calculation with the decimal place–value system are already systematically presented in al-Khwārizmī's very influential book on arithmetic (written in the ninth century), which only

---

[10] See Bobzien 2006: sect. 5.   [11] See MacFarlane 2000; Dutilh Novaes 2011c; Dutilh Novaes 2012.
[12] See Bellos 2010: chap. 3. It has been argued (Lam 1996) that the Indian place–value system was in fact borrowed from Chinese mathematics, which also had a decimal place–value system evolved from the practice of using counting rods for calculation. Chinese mathematics will not be discussed in any detail in the remainder – not because it is not interesting, but for reasons of space.

survives in Latin translations and was usually referred to by its first words in the Middle Ages: *Dixit algorismi*, 'Thus spoke al-Khwārizmī'.[13] Its original title was probably 'The Book of Addition and Subtraction According to the Hindu Calculation', as al-Khwārizmī was indeed for the most part relying on techniques developed by Indian mathematicians.[14]

What is most significant for our purposes is that the place–value system as a *notational* device was to have tremendous cognitive impact on the development of mathematics. This system later allowed for the formulation of calculating techniques which would significantly facilitate numerical calculation. In the absence of a notational system which serves the purpose not only of *representing* quantities but also of providing a medium to be *operated on*, traditional methods of calculation relied on instruments such as the abacus or counting rods. The very idea that one can use portions of writing to operate on just as one would use an instrument such as an abacus or counting rods is a decisive step towards the development of symbolic systems for reasoning.[15]

Further on, the main elements in the (proto-)history of formal languages are the notational developments for algebra,[16] so in what follows I present a highly condensed account thereof.[17] These developments are usually described in terms of three main stages (following G. H. F. Nesselmann):

(a) 'rhetorical' algebra, in which everything is set out in full words; (b) 'syncopated' algebra, in which standardized abbreviations or signs are used, but the stenographic expression still represents language; (c) 'symbolic algebra', in which ... operations are performed directly on [the symbolic] level. (Høyrup 2006: 1–2)[18]

There are two authors to whom the honorific 'father of algebra' is usually attributed: the third-century Greek mathematician Diophantus of Alexandria and the (aforementioned) ninth-century Persian mathematician al-Khwārizmī. Diophantus' major work is his *De Arithmetica* treatise, of which only six (out of thirteen) books survive. It is a collection of problems

---

[13] As already mentioned in Chapter 1, this is the origin of the term 'algorithm', the Latin rendition of 'al-Khwārizmī' for the title of this text.

[14] See Høyrup 2006.

[15] Interestingly, some techniques and notations developed within Chinese algebra clearly emerged from the practice of calculating with counting rods, illustrating the tight connection between calculating devices and certain notational systems (De Cruz 2007).

[16] Again, with the caveat that, for reasons of space, I do not discuss the influence of the Lullist tradition at length.

[17] Naturally, some important elements will be left out for reasons of space, but I believe the presentation here contains the main steps.

[18] Notice, however, that the fruitfulness of this tripartite distinction has been questioned in Heeffer 2007.

receiving numerical solutions by means of equations, but it offers no systematic account of general methods to be used in a wider range of cases. It differs, however, from previous texts in that the problems are not formulated in terms of cattle, grains, etc., but rather in purely numerical terms. The received view (initially promoted by Nesselmann) is that Diophantus was the first to introduce a symbolic representation (the Greek letter sigma) for the *unknown value*, i.e., the information to be disclosed by means of a calculation.[19] However, given that the oldest extant manuscript of this text dates to the thirteenth century, we cannot be sure that the symbol for the unknown value was not a much later scribal addition.[20] Thus, Diophantus' status as the 'father of algebra' is questionable.

By contrast, al-Khwārizmī's 'Book on restoration and opposition' presents general techniques, and this is why it is usually viewed as the founding text of algebra (although al-Khwārizmī was in all likelihood offering a compendium of previously available techniques). Crucially, though, al-Khwārizmī's algebra is purely *rhetorical*: no use is made of specific symbolism, not even abbreviations. It is also important to notice that al-Khwārizmī's demonstrations in the treatise of the six rules of algebra (the only demonstrations in the treatise, technically speaking) are all geometric proofs, in the traditional Euclidean style (Høyrup 2010).

The road from rhetorical to symbolic algebra was long and winding. It is widely agreed that it is only with Viète and especially Descartes that we can speak of fully fledged symbolic algebra, and previous developments are still the object of study and controversy among scholars. But some scholars argue for the compelling thesis that the first introductions of special symbols (yielding what is referred to as 'syncopated algebra') occurred in the Maghrebian algebraic tradition, including the development of notations for fractions (Høyrup 2010: sect. 1.1). A curious characteristic of these early uses of symbolic notation in Maghrebian texts is that they are usually not incorporated in the main text, and often correspond to something that is also stated rhetorically. Clearly, the authors were cautiously experimenting with a new notational technology, and resorted to more familiar modes of expression to anchor the exploration.

---

[19] Recall that schematic letters had been in use in Greek logic and geometry for many centuries, and Diophantus' introduction of a symbol for the unknown value seems to have no direct relation with the use of schematic letters elsewhere. This is significant in that it outlines the historical independency of the notions of a schematic letter and of a 'variable' (insofar as denoting the unknown quantity can be seen as the precursor of the concept of a variable), which are often conflated.

[20] See Heeffer 2007; Netz 2012.

Though manuscripts differ in this respect ... the symbolic calculations appear to have been often made separate from the running text ... usually preceded by the expression 'its image is'. They illustrate and duplicate the expressions used by words. They may also stand as marginal commentaries. (Høyrup 2010: 9)

The Maghrebian tradition seems to have had considerable influence on the mathematics taught in the medieval abbaco schools, a sub-scientific tradition which was of great significance in the development of mathematics in Europe (Høyrup 2010). Al-Khwārizmī's two major books (on algebra and on arithmetic) had been translated into Latin already in the twelfth century, but the academic curriculum remained largely dominated by the established approach to mathematics. Here is an apt description of the situation:

By the end of the fifteenth century there existed two independent traditions of mathematical practice. On the one hand there was the Latin tradition as taught at the early universities and monastery schools in the quadrivium. Of these four disciplines arithmetic was the dominant one with *De Institutione Arithmetica* of Boethius as the authoritative text. Arithmetic developed into a theory of proportions as a kind of qualitative arithmetic rather than being of any practical use, which appealed to esthetic and intellectual aspirations. On the other hand, the south of Europe also knew a flourishing tradition of what Jens Høyrup (1994) calls 'sub-scientific mathematical practice'. Sons of merchants and artisans, including well-known names such as Dante Alighieri and Leonardo da Vinci, were taught the basics of reckoning and arithmetic in the so-called abbaco schools in the cities of North Italy, the Provence, and Catalonia. The teachers or *maestri d'abbaco* produced between 1300 and 1500 about 250 extant treatises on arithmetic, algebra, practical geometry and business problems in the vernacular. The mathematical practice of these abbaco schools had clear practical use and supported the growing commercialization of European cities. These two traditions, with their own methodological and epistemic principles, existed completely separately. (Heeffer 2007)

Indeed, modern algebra (and its notation) will ultimately emerge from the sub-scientific tradition of the abbaco schools, rather than from the somewhat solidified academic tradition taught at the medieval universities. Høyrup (2006) and Heeffer (2010) offer an account of the immediate predecessors of Viète's symbolic algebra (who were, broadly speaking, relying on the work of the *maestri d'abbaco*), in particular of 'the transformation of algebra from rules for problem solving to the study of equations' (Heeffer 2010: 57).

The 'birth of modern mathematics' with Viète and Descartes has been extensively analysed by a large number of scholars,[21] and the main point for our purposes is that the transformations introduced by both are inherently

---

[21] The main reference here is Serfati 2005.

connected to *notational* innovations. Viète's main innovation was the introduction of two classes of symbols: uppercase vowels for unknown values, and uppercase consonants for the known parameters of a problem. This was an absolutely essential step in the development of modern algebra, and more generally in the development of the technology of symbolic systems. Viète claims to have been directly influenced by Diophantus' *De Arithmetica* (which had been newly translated) and its introduction of symbols for the unknown and for powers of the unknown (Macbeth 2004), but it is clear that he was also greatly indebted to the algebraic tradition ranging from al-Khwārizmī, the Maghrebian tradition, and the Latin abbaco schools as well (Heeffer 2010; 2007). Naturally, with Viète's innovations, the theoretical, abstract study of equations was greatly facilitated, as the symbols for known parameters allowed for the study of *classes* of equations.

Nevertheless, Viète's way of writing mathematics still strikes the modern reader as rather 'rhetorical'; indeed, the final step in the establishment of mathematical notation as we know it (in particular concerning equations) was taken by Descartes, as dramatically illustrated by the following passage:[22]

The equation that Luca Pacioli in 1494 would have expressed as:
4 Census p 3 de 5 rebus ae 0
and Viète would have written in 1591 as:
4 in A quad – 5 in A plano + 3 aequatur 0
in 1637 Descartes had nailed as:
$4x^2 - 5 + 3 = 0$
(Bellos 2010: 182)

Descartes introduced the convention, still in use today, of denoting known quantities with the first letters and unknown quantities with the last letters of the alphabet (all lowercase, differing from Viète).[23] It is indeed this 'Cartesian' notational background and the spectacular advancement of mathematics in the seventeenth century that will serve as inspiration for the first systematic discussions of formalisms as general methods of inquiry in the seventeenth century, with Leibniz in particular. Thus, in the seventeenth century, the idea that portions of writing can be powerful calculation tools (just as abacuses or counting rods) was fully established, with

[22] Notice that Pacioli would most likely not have expressed this exact equation, and moreover that he wrote in Italian, not in Latin. Nevertheless, for the purposes of illustration, this passage is (although inaccurate) quite striking.
[23] Recall Landy and Goldstone's remark at the start of this chapter on how such conventions help offloading working memory.

algorithms for calculation with numbers and the expression of equations in symbolic form accompanied by well-defined ('syntactic') rules of transformation.[24]

So far in our story, the novel notational technologies are still restricted to mathematics and its applications, while in logic no significant new developments have been introduced: regimentation and schematic letters remained the main tools. How and when were the new notational technologies brought over to logic? The key factor seems to have been a remarkable innovation, starting in the sixteenth century and consolidated in the seventeenth century: 'the attempt to connect the two previously separated disciplines of logic and mathematics' (Mugnai 2010: 297). While there had been different instances of interplay between logic and mathematics prior to this, these had been occasional events; systematically, the two disciplines were viewed as radically dissimilar, as attested by the trivium/quadrivium separation in the scholastic curriculum. Here is Mugnai's informative summary of the seventeenth-century developments:

This happens along two opposite directions: the one aiming to base mathematical proofs on traditional (Aristotelian) logic; the other attempting to reduce logic to a mathematical (algebraic) calculus. This second trend was reinforced by the claim, mainly propagated by Hobbes, that the activity of thinking was the same as that of performing an arithmetical calculus.

Naturally, the second direction is particularly relevant for our purposes, as it represents a straightforward channel for the importation of mathematical notational techniques into logic.[25] But before we examine these developments more closely a few philosophical observations are in order. Notice that, traditionally, the logical object/item is the *argument* connecting premises to conclusions, as already formulated in Aristotle's famous definition of a syllogism/deduction:

A deduction is a discourse in which, certain things having been supposed, something different from the things supposed results of necessity because these things are so. – *Prior Analytics* 24b19–23. (Smith 1989)

Thus, an argument is a *discourse*, and its main components are the premises and the conclusion; it is deductively valid if the truth of the premises necessitates the truth of the conclusion. This template applies to deductive

---

[24] 'The symbolic equation slowly emerged during the course of the sixteenth century as a new mathematical concept as well as a mathematical object on which new operations were made possible' (Heeffer 2010: 57).

[25] For reasons of space, nothing will be said about the first direction; the interested reader should consult Mugnai 2010.

arguments more generally, including geometrical proofs in the style of Euclid's *Elements*. Now, besides this characteristic style of argumentation, a different kind of mathematical practice was even more widespread, both historically and geographically: the practice of finding the (unique) solution to a problem, given some initial known parameters. This practice is closer to the traditional practical applications of mathematics such as those related to bookkeeping and measuring, and can be referred to as the practice of *calculation*, as opposed to the practice of *demonstration*. Naturally, this distinction is a simplification, and the work of Chemla (2005) on ancient Chinese mathematics has shown that Chinese mathematicians, for example, also offered *proofs* of the correctness of their calculating procedures. Nevertheless, the point of drawing this distinction is to illustrate the importance of these two practices coming together in the 'marriage' between logic and algebraic mathematics in the seventeenth century.

At first sight, there seems to be a neat analogy between the premises and the conclusion of an argument, on the one hand, and the known parameters and the unknown value in a calculation, on the other hand, but ultimately these are very different concepts. In particular, proving a theorem does not seem to have the same dimension of *discovery* from the known to the unknown: the conclusion is typically already known, and the point is rather to show that the truth of the conclusion follows from the truth of the premises (for example, to compel the interlocutor to accept the conclusion if she grants the premises).[26] In a slogan, an argument, proof, or demonstration is a *discourse*; a calculation is a *procedure*.

While calculating practices and techniques are ubiquitous in all mathematical traditions, the corresponding level of theoretical maturity and generality attained by ancient Greek geometry is arguably only reached with the advent of what we refer to as 'algebra', i.e., a theory of equations. As we have seen, and following Høyrup (2006), there are already traces of what can be considered as (proto-)algebra in Diophantus, in ancient Chinese mathematics, and in Indian mathematics, but the establishment of algebra as a discipline is generally speaking a more recent event. Indeed, recall that al-Khwārizmī resorts to *geometrical proofs* in his algebraic work, thus relying on a more solidly established canon of argumentation and justification. Hence, the seventeenth-century trend towards a reduction of logic to algebraic calculus is a remarkable and non-trivial event in the history of

---

[26] In fact, the unknown value in a proof/demonstration is the proof itself, i.e., the argumentative path going from premises to conclusion (I owe this point to Luis Carlos Pereira).

logic and mathematics (and their relations): it brings together two bodies of practices that are historically and conceptually divergent.

The general idea of an 'algebraization' of logic in the seventeenth century is most readily associated with Leibniz, but, as argued by Mugnai (2010), it was a general trend pursued by several people, often independently of each other.[27] The main factors behind this development appear to have been the astonishing progress in algebra with Viète and Descartes,[28] and the Hobbesian idea, which in fact can be traced back to the influential Ramon Lull, that thinking in general amounts to performing calculations.

As is well known, Lull designed 'machines' that he thought could convert the infidel to Christianity by showing him the infinite perfections of the Christian God (Lohr 2010). Apart from this rather fantastic goal, the general approach contained some radically novel ideas:

> One of the great ideas implicit in Lull's work is that *reasoning can be done by a mechanical process*. Another equally profound idea in Lull's thought is that *reasoning does not proceed by syllogism but by combinatorics. Reasoning is the decomposition and recombination of representations*. The decomposition and recombination of attributes can be represented by the decomposition and recombination of *symbols*, and that, as Lull's devices illustrate, is a process that can be carried out by machines. (Glymour 1997: 71)

Lull's ideas remained influential in the later Middle Ages and the Renaissance (Lohr 1974), even though this influence is not felt in many of the 'canonical' authors of these periods (but his influence over Nicholas of Cusa, for example, has been well documented). Arguably, Lullism forms the historical background (even if only indirectly) for Hobbes's conception of thinking as computation. In his *Computatio*, Hobbes writes:[29]

> By ratiocination, I mean computation. Now, to compute is either to collect the sum of many things that are added together, or to know what remains when one thing is taken out of another. Ratiocination, therefore, is the same with addition and subtraction. (Hobbes 1839: 3)

---

[27] Krämer 1991 provides the definitive account of these developments.

[28] 'The art of combinations ... signifies purely the science of forms or formulas, or even of variations in general. In a word, it is the universal specious arithmetic or characteristic ... One can even say that calculation with letters, or more precisely algebra, is in a certain sense subordinate to it, because one employs many signs which are indifferent ... For this aim the letters of the alphabet are highly suitable. And when these letters or signs signify magnitudes or numbers in general, the result is algebra or rather Viète's specious arithmetic ... When these letters signify terms or concepts, as in Aristotle, this gives us that part of logic which deals with figures and modes': Leibniz, 'Sur la calculabilité du nombre de toutes les connaissances possibles' (Mugnai 2010: 309).

[29] Again, it must be borne in mind that this was a revolutionary statement at the time, even though it now sounds somewhat naive.

The cornerstone of the process of algebraization of logic was the establishment of correspondences between traditional logical concepts, such as the
affirmative and the negative copulas, and algebraic operations, as described
in this passage:

That most profound investigator of the principles of all things, Thomas Hobbes,
has rightly contended that every work of the human mind consists in computation,
and on this understanding, that it is effected either by adding up a sum or
subtracting a difference ... Accordingly, just as there are two primitive signs
used by Algebraists and Analysts, + and –, so there are, as it were, two copulae, *is*
and *is not*: with the former the mind puts things together, with the latter it takes
them apart.[30] (Leibniz 1966: 194)

At the time, many thinkers were toying with the idea of an algebraization of
logic, proposing different mappings of logical concepts into algebraic operations. A paradigmatic text is Jacob Bernoulli's *Parallelism between Logical
and Algebraic Reasoning*, where the author establishes, among others, the
correspondence between the particle 'and' and the operation of addition,
represented by '+' (Mugnai 2010: sect. 5). Leibniz's own main contribution
was arguably the idea that the computational operations of the mind – in
fact, thinking in general – correspond first and foremost to the manipulation of *signs*.[31]

As is well known, Leibniz contemplated the ambitious goal of reducing all
human thought to a few primitive concepts, from which all knowledge could
be derived/produced by 'mechanical' means, once signs would be assigned to
each primitive concept. In this context, he introduced a conceptual distinction that remains relevant for contemporary discussions: the distinction
between *calculus ratiocinator* and *lingua rationalis*, which was then extensively
used by van Heijenoort in his historiography of logic – yielding the 'logic as
language v. logic as calculus' dichotomy (Peckhaus 2004).

Leibniz used the term *characteristica universalis* to refer to the collection
of symbols which would (presumably) cover all simple/primitive thoughts
(possibly also with grammatical rules on how to combine them); in modern

---

[30] Naturally, thus formulated, the parallelism sounds rather crude. It is only with Boole that such
systematic correspondences were more carefully developed.
[31] 'A typical piece is the short tract "Fundamenta calculi ratiocinatoris" written presumably during
[Leibniz's] stay in Vienna between May 1688 and February 1689. In this text, Leibniz stresses that all
human reasoning is connected to some signs or characters. Characters are signs perceptible with the
senses, e.g., being written down, or cut into stone. But such abbreviating signs should not only be
applied to the things themselves, but also to the ideas of things. "Abbreviating" means that as soon as a
characteristic sign has been established for a complex object, memory can be relieved of the burden of
retaining all the characteristic elements of this object' (Peckhaus 2004: 7).

terms, it would roughly correspond to the syntax of a (logical) language, which defines its basic terms and has strict rules of formation determining what counts as a well-formed formula. The *calculus ratiocinator*, in turn, corresponds to the calculating procedures to be employed so that new truths can be deduced from known ones in a 'mechanical' way; again, in modern terms, it would correspond to the rules of transformation of a calculus, the rules of inference in a logical system being a special case thereof. For Leibniz, a *calculus ratiocinator* together with a *characteristica universalis* would form the *lingua rationalis*, with which knowledge could be attained (Leibniz 2000); critically, *both* aspects are equally important. For our purposes here, one should note that, while in general a formal language is currently viewed as roughly corresponding to what Leibniz describes as *characteristica universalis*, i.e., pertaining to the vocabulary and the rules of formation alone, it is only once appropriate rules of transformation are added to a language that it becomes a proper tool for reasoning. Thus, 'formal languages' are here understood in a broader sense (following Carnap), covering both the vocabulary/rules of formation and the rules of transformation – a *lingua rationalis*.[32]

Another noteworthy aspect of Leibniz's project of mathematization of logic and reasoning was his multiple interpretations of Aristotelian syllogistic as numerical calculi. He was looking for methods which would reduce the establishment of the validity of an argument to a simple calculation, on the basis of his notion of 'characteristic numbers'.[33] The general idea of 'calculating' the validity of arguments would later be pursued by Boole and others in the algebra of logic tradition.

Thus, the seventeenth century saw the birth of an algebraic approach to logic and reasoning, most notably but not exclusively exemplified by Leibniz. For our purposes, the main point is that the idea of using symbolic systems such as mathematical notations or even specially designed formalisms in *logical* investigations became widely accepted, even if not thoroughly implemented.[34] The approach remained alive in the eighteenth century (Mugnai 2010: sect. 7), but its next truly remarkable instantiation had to wait for the nineteenth century, when Boole's work gave a whole new momentum to the project.

[32] In modern terminology, the term 'formal system' is typically used.
[33] See *On Characteristic Numbers* (Leibniz 1989).
[34] As is well known, the general idea remained essentially programmatic until Boole's work. Besides the exchanges between logic and mathematics, the seventeenth century is also marked by a strong interest in so-called 'artificial languages', generally speaking, with intended applications going well beyond the realm of these two disciplines. Here again Leibniz is a central figure, but other names to be mentioned are Wilkins and Dalgano (Maat 2004).

3.1.2 The era of formal languages

The eighteenth century was not a period of major innovations in logic, but great discoveries were made in different areas of mathematics. In algebra in particular, Lagrange and Laplace (among others) introduced new and powerful techniques. Thus, the algebra of the first half of the nineteenth century was very different from the algebra of the seventeenth century, which had offered the background for the first attempts at a 'mathematization' of logic. George Boole would be the first to apply this cutting-edge mathematics to an analysis of logic.

Boole was a self-taught mathematician, who at the beginning of his career combined mathematical research (differential equations, integration, and the calculus of variation) with his work as a schoolmaster. For most of his career, he focused indeed on standard topics in mathematics; it is only in the period between his two logic books, *The Mathematical Analysis of Logic* (1847) and *The Laws of Thought* (1854), that he worked predominantly on logic, more specifically on an algebraic approach to logic. Boole not only contemplated the possibility of establishing mathematical counterparts to traditional logical concepts, as had been done by seventeenth-century thinkers: he in fact developed a system where the traditional categorical propositions could be expressed by means of *equations*, so that algebraic operations and techniques could be applied. Symbols would represent classes, and transformation rules borrowed from algebra would be applied to these 'logical' equations.

Just as Leibniz had done in the seventeenth century with characteristic numbers, Boole proposed a calculus to account for syllogistic validity, but this time formulated in purely algebraic terms. Here's the main idea:

[Boole's] first observation was that syllogistic reasoning was just an exercise in *elimination*, namely the middle term was eliminated to give the conclusion. Elimination was well known in the ordinary algebraic theory of equations, so Boole simply borrowed a standard result to use in his algebra of logic. If the premises of a syllogism involved the classes X, Y, and Z, and one wanted to eliminate y, then Boole put the equations for the two premises in the form:

$$ay + b = 0$$

$$a'y + b' = 0.$$

The result of eliminating y in ordinary algebra gave the equation

$$ab' - a'b = 0,$$

and this is what Boole used in his algebra of logic to derive the conclusion equation. (Burris 2010: sect. 3)

There were some complications which needed to be dealt with for this technique to be generally applicable, but the main point is that Boole approached logic solely through equations, which allowed him to import results from algebra; he thus founded the tradition of the 'algebra of logic'. While still not a fully fledged formal language, with Boole (for the first time) we have a symbolic system with strict rules of formation and transformations being used for logical investigations. Accordingly, many historians view Boole's *The Laws of Thought* as marking the birth of mathematical logic, even though his system is still riddled with short-comings (Corcoran 2003).

One might think that Boole (and others in the same period, such as De Morgan) may have been influenced by the Leibnizian project of mathematizing logic, but in fact it seems that there were no direct channels of influence between Leibniz and Boole (as is well known, most of Leibniz's logical writings were only published centuries after their composition).[35] As well described by Peckhaus in the title of his monograph (1997), an algebraic approach to logic was *rediscovered* in the nineteenth century; indeed, with the continuous advancement of algebra, the structural analogies between algebra and logic became increasingly manifest once more.

Thus, Boole inaugurated a new tradition (Burris 2009), which offered a most fruitful background for work in logic in the nineteenth century, and had (after Boole) Jevons, Peirce, and Schröder as its main figures. The key idea in this tradition was that of providing an *algorithmic treatment* to logical analysis, by means of the importation of symbolism and techniques from algebra. In practice, their theories resulted in very general approaches, allowing for, for example, arguments with more than two premises, which Aristotelian syllogistic cannot accommodate.

Besides his technical contributions to the algebra of logic, Peirce is also known for his introduction of non-linear iconic systems of representation for logical analysis; in particular, his Existential Graph (EG) is formally equivalent to a (sentential) predicate language. Diagrams had been used in connection with logic (syllogistic reasoning in particular) by a number of authors, such as Euler, Venn, and Carroll. But Peirce's EG goes much

---

[35] 'No doubt, the new logic emerging in the second half of the 19th century was created in a Leibnizian spirit. The essentials of Leibniz's logical and metaphysical programme and of his idea concerning a logical calculus were available at least since the 1840s ... As soon as these logicians became aware of Leibniz's ideas, they recognized Leibniz's congenial affinity and accepted his priority. But the logical systems had basically been already established. Therefore there was no initial influence of Leibniz on the emergence of modern logic in the second half of the 19th century' (Peckhaus 2009: sect. 6).

beyond these in that it contains a full syntax defining transformations on the 'formulae' of the language (diagrams) themselves, i.e., rules of transformation for a non-sentential system (Legg 2011). Peirce's work was, however, too visionary to be understood and assimilated into mainstream logic, and it is only in recent decades that the significance of diagrammatic reasoning and multi-modal representational systems has been given more attention, in the work of Barwise, Etchemendy, Shin, and others (Shin and Lemon 2001).

Now, while all historians acknowledge the innovative character of Boole's work and the sophistication of the algebra of logic tradition, most authors (following van Heijenoort) still date the birth of modern logic with Frege's *Begriffsschrift*, in particular due to the fact that it is the first system with a fully fledged formal language to be used for logical investigations. Frege introduces a completely novel two-dimensional (i.e., non-linear) symbolic system, which he claims to be much better suited for the expression of mathematical theorems than the usual languages of mathematics. Thus, the first point of dissimilarity with respect to Boole is that, while Boole imported notations familiar from algebra, Frege essentially designed a new language from scratch, which contained just a few familiar symbols (in particular the use of letters to indicate generality, which had been in use since Viète).[36]

The second point of dissimilarity is that, while Boole seems not to have been aware of Leibniz's writings at all when first developing his system, Frege overtly claims to be inspired by Leibniz's *ars characteristica* project. (In turn, it seems that Frege was at first not familiar with Boole's work, and only became acquainted with it when he had to respond to criticism put forward by Schröder.)

The *Begriffsschrift* remains inspiring but puzzling. I will have much more to say on Frege's 'ideography' (in the terminology of one of its English translations) shortly, when discussing the different avowed purposes of using a formal language, but for now the two main points worth noticing are the fact that it contains a precisely defined syntax, both for rules of formation and rules of transformation, but also that its two-dimensional, non-linear notation was never widely adopted. Thus, in the broader picture

---

[36] 'The most immediate point of contact between my formula language and that of arithmetic is the way in which letters are employed' (Frege 1879: 6). 'I adopt this basic idea of distinguishing two kinds of signs, which unfortunately is not strictly observed in the theory of magnitude, in order to apply it in the more comprehensive domain of pure thought in general. I therefore divide all signs that I use into those by which we may understand different objects and those that have a completely determined meaning. The former are letters and they will serve chiefly to express generality' (Frege 1879: 10–11).

of the history of formal languages sketched here, its main contribution concerns the very idea of using formal languages (in the sense of a language with strict rules of formation and transformation, i.e., in the sense of the formal as computable) for logical investigations, but not the adoption of particular notational conventions.[37]

In effect, as is well known, Frege's system did not have a particularly significant impact in the years following the publication of *Begriffsschrift*, and was heavily criticized by Schröder in particular (Peckhaus 2004: sect. 4); nobody seems to have felt compelled to adopt it. It was only with Whitehead and Russell's *Principia Mathematica* that *Begriffsschrift* became widely influential, but in an indirect manner. Whitehead and Russell claimed to be following Frege's footsteps and to be adopting his innovations, but in practice (and perhaps unbeknownst to them) the system in *Principia* differs considerably from Frege's ideography. In particular, they abandon Frege's two-dimensional notation in favour of Peano's linear, sentential notation; while this move may have seemed quite innocent from their point of view, it betrays a serious misunderstanding of the exact purpose and force of the *Begriffsschrift* system (as will be discussed shortly).

Principia Mathematica represents the first influential logical system where rules of inference (rules of transformation) are treated with the same rigour as the formulation of axioms. As narrated by Awodey and Reck (2002), in the last decades of the nineteenth century, the programme of axiomatizing portions of then-contemporary mathematics was in full swing, but for a number of years those involved in the programme did not seem to realize that reliance on an intuitive, non-formalized approach to inference/deductive consequence was an important limitation.[38] Without a rigorously defined system of rules of inferences, it is nearly impossible to tackle meta-properties of a given axiomatic system (completeness, in particular) in a technical (as opposed to conceptual) manner. While *Principia* itself does not seem to display a clear notion of meta-level investigation, it seems to have provided a crucial ingredient for the

---

[37] It is also worth noticing that many mathematicians working in the second half of the nineteenth century expressly rejected the 'calculative' approach to mathematics which had prevailed in the seventeenth and eighteenth centuries (the golden era of algebra) (see Macbeth, 'Writing Reason' (forthcoming)). Instead, they emphasized conceptual analysis and foundational investigations, and this is the background for the programme of axiomatizing portions of mathematics (e.g., Dedekind).

[38] 'From a contemporary point of view the main ingredient missing in the works considered so far is a precise and purely formal notion of deductive consequence' (Awodey and Reck 2002: 19).

development of meta-theoretical analyses.[39] Indeed, in the first decades of the twentieth century, logicians and mathematicians became increasingly aware of the fact that desirable meta-properties such as completeness (in different senses) could not simply be taken for granted; they would not necessarily follow from a well-crafted axiomatization (completeness is treated further in section 3.3.3).

Besides the gradual adoption of a rigorous approach to rules of inference (rules of transformation) – which, I claim, was for the most part motivated by the need for meta-theoretical analyses – another key, early twentieth-century development towards modern conceptions of formal languages was the idea that a given symbolic system can be uninterpreted or reinterpreted. Essentially, it corresponds to what I have described as 'the formal as de-semantification' in Chapter 1, and while a general move towards de-semantification can be identified in many of the previous developments in mathematical notation (going as far back as the abbaco schools tradition), it is only at this point that de-semantification becomes conceptualized as such in the domain of logic.[40]

The main figure in this respect is, of course, David Hilbert, who allegedly once claimed that 'one must be able to say "tables, chairs, beer mugs" each time in place of "points, lines, planes"'. Grattan-Guiness (2000: 208) observes that 'this famous remark is normally misunderstood and Hilbert may not have thought it through at the time', but the gist of it seems to be a defence of what can be (anachronistically) described as a 'model-theoretic' approach, one where intuitions concerning the objects in question have no role to play. Instead, the emphasis is laid on the logical relations between concepts and axioms (independence in particular) in such a way that the system should be interpretable onto alternative collections of objects.[41] The concept of de-semantification, and Hilbert's technique of *reinterpretation* in particular, is discussed in more detail in Chapter 6, but for now it is important to flag it as an important step in the development of formal languages in the early twentieth century.

Later on, in the 1920s, Hilbert put forward a new approach to the foundations of mathematics, calling for a full formalization of mathematics

[39] 'Although the authors of *Principia* did not cast their logic into a formal axiomatic mold ... they did convince several mathematicians and logicians of the value of their new, more formal approach to logical deduction, notably Hilbert and Rudolf Carnap' (Awodey and Reck 2002: 19–20).
[40] As we have seen in Chapter 1, some nineteenth-century authors such as Thomae already advocated a 'formal' approach to mathematics (arithmetic in particular), but Hilbert's approach is nothing like Thomae's 'marks-on-paper formalism'.
[41] See Hilbert's 'Grundlagen der Geometrie' (1899).

in axiomatic form so that meta-theoretical questions – in particular consistency – could be investigated from a purely mathematical point of view (Zach 2003). The upshot was that these axiomatizations (including the underlying language and rules of transformations) were no longer seen exclusively as being *about* mathematical objects (number, point, line, etc.); they became mathematical objects *themselves*. With respect to Hilbert's earlier work, one of the main differences was that, post-*Principia Mathematica*, which exerted great influence on him (Zach 2003: sect. 1.2), an axiomatic system was now defined not only by rigorously formulated axioms but also by strict rules of transformation within the system. Thus formulated, an axiomatic system could be viewed as a fully fledged mathematical object, which could then be studied by means of mathematical techniques.

Essentially as a result of the meta-theoretical turn promoted by Hilbert's programme, by the 1930s formal languages were not only systems *with which* one could prove theorems (as had been Frege's intention with *Begriffsschrift*), they were also (and perhaps more centrally) systems *about which* one could prove theorems. This is an absolutely crucial transformation, which leaves behind the traditional role assigned to notations in the development of mathematics as calculating tools, i.e., as objects *with which* to reason (both for expressive and calculative purposes).

Still in the same spirit, but motivated by different considerations, in the 1930s, Tarski introduced the concept of a hierarchy of languages (Tarski 1944). In order to block the emergence of Liar-like paradoxes, in his 'semantic' approach to truth, Tarski suggested that a given theory should not contain its own truth-predicate; instead, the truth-predicate of a language L_1 would only be expressible in a language L_2, whose own truth-predicate would be expressible in a higher-order language L_3, and so forth. Thus, with Tarski, the object-level v. meta-level distinction itself became formalized.

In the aftermath of Hilbert's programme, even after the blow inflicted by Gödel's incompleteness theorems (cf. section 3.3.3), the metalogical perspective clearly gained the upper hand, almost entirely surpassing the conception of logic as a tool for reasoning: logical systems are not to reason *with*, but rather to reason *about*. This holds both of the model-theoretic and of the proof-theoretic traditions. In the former, one is chiefly interested in the soundness and completeness of a given deductive system with respect to a semantics. In the latter, the meta-properties under investigation are normalization, cut-elimination, the sub-formula property, etc. Currently, the meta-theoretical perspective is still predominant, even though there are some important uses of logical systems as tools to investigate specific phenomena, in what is now known as the domain of applied logic (the pure v. applied distinction regarding logic is further treated in Chapter 7).

And what happened to the age-old logical techniques of regimenting and using schematic letters? Have they been entirely supplanted by the imported novelties from mathematics? An inspection of cutting-edge current research in logic might suggest that they have, but this is in fact not the case. These techniques are alive and kicking in the doctrine of the 'logical form' of sentences and arguments, and are particularly conspicuous in introductory logic textbooks (especially those written for philosophy students).[42] Here is an example:

We say that (1), (7) and (8) [examples of arguments previously given] have a particular *form* in common, and that it is this form which is responsible for their validity. This common form may be represented schematically like this:

> (11)   A or B
>
>   Not A
>   -----------
>   B

These schematic representations are called *argument schemata*. The letters A and B stand for arbitrary sentences. Filling in actual sentences for them, we obtain an actual argument. Any such substitution into schema (11) results in a valid argument, which is why (11) is said to be a valid argument schema. (Gamut 1991: 3)

A little further:

Logic, as the science of reasoning, investigates the validity of arguments by investigating the validity of schemata. For argument schemata are abstractions which remove all those elements of concrete arguments which have no bearing on their validity. (Gamut 1991: 4)

Such claims can be found in virtually every introductory textbook to logic, and are thus the kind of 'philosophical' accounts of logic that a student is most likely to encounter in her/his first contact with logic.[43] Elsewhere (Dutilh Novaes 2012), I refer to this doctrine as LHAWKI, 'logical hylomorphism as we know it', which is highly problematic for a variety of reasons. Nevertheless, it is beyond discussion that the techniques of

---

[42] They also permeate the project of formal semantics, which came into existence with Montague's idea of applying the usual logical apparatus, which was supposed to represent an improvement over ordinary languages (see next section), to the study of ordinary languages themselves.

[43] Hodges (2009: 592) correctly notes of Aristotle's regimentation practice described in 3.1.1, above: 'So far, his practice seems almost identical with what we do today in elementary logic classes.'

regimentation and the use of schematic letters have been crucial (also in a positive sense) for the development of logic as a discipline through the centuries.

3.2 WHAT ARE FORMAL LANGUAGES GOOD FOR?

In the previous section, I presented a survey of the main steps in the historical development of formal languages, but did not extensively discuss the *motivations* given for the adoption of this new technology. What is the rationale behind developing and adopting formal languages? What will be the pay-off of the investment in learning the new technique? We now tend to take the use of formal languages in logic for granted, and thus tend not to ask ourselves why we do it at all. However, the formal languages pioneers spent a great deal of time justifying their uses, essentially on the basis of three main functions: expressivity, depiction, and calculation.

3.2.1 *Expressivity*

Perhaps the most frequent and most deeply engrained justification for the adoption of formal languages is the (presumed) inadequacy of previously available means of expression and communication, in particular what is (misleadingly) referred to as 'natural language', in a given domain or for a given task. The inadequacy is usually attributed to the 'imperfections' of ordinary languages such as lack of precision and perspicuity, the presence of ambiguities and other deficiencies. From this point of view, at least three positions can be (and have been) maintained (Stokhof 2007): formal languages simply *extend* but do not replace ordinary languages; formal languages represent an *improvement* in that they allow us to perform certain tasks with more ease, but which could at least in theory be performed without them; formal languages are meant to *reform* ordinary languages, which are so hopelessly inadequate that they must be thoroughly sanitized.

The *locus classicus* (at least in the recent past) for the idea that formal languages come to remedy shortcomings in ordinary languages is, of course, the preface of the *Begriffsschrift*. Frege's general project is that of providing logical foundations for mathematics, and for this purpose one must spell out mathematical proofs in an entirely perspicuous way, outlining all inferential steps and all assumptions made. It is thus his *logicism* which (at least in part) motivates the formulation of such a language, so that the purely logical can be isolated from the non-logical (intuitive) in mathematical proofs. In a much-quoted passage from the preface of *Begriffsschrift*, Frege states his motivations and goals very clearly:

My first step was to attempt to reduce the concept of ordering in a sequence to that of *logical* consequence, so as to proceed from there to the concept of number. To prevent anything intuitive from penetrating here unnoticed, I had to bend every effort to keep the chain of inferences free of gaps. In attempting to comply with this requirement in the strictest possible way I found the inadequacy of language to be an obstacle; no matter how unwieldy the expressions I was ready to accept, I was less and less able, as the relations became more and more complex, to attain the precision that my purpose required. This deficiency led me to the idea of the present ideography. Its first purpose, therefore, is to provide us with the most reliable test of the validity of a chain of inferences and to point out every presupposition that tries to sneak in unnoticed, so that its origin can be investigated. (Frege 1879: 5–6)

We will see later that Frege's ideography in practice also has a 'hands-on', operative dimension, but his initial and primary motivation seems to have been a concern regarding the inadequacy of ordinary languages for the logicist project.

In another seminal text, *The Logical Syntax of Language*, Carnap makes essentially the same point, but this time not in the same logicist spirit:

In consequence of the unsystematic and logically imperfect structure of the natural word-languages (such as German or Latin), the *statement* of their formal rules of formation and transformation would be so complicated that it would hardly be feasible in practice. And the same difficulty would arise in the case of the artificial word-languages (such as Esperanto); for, even though they avoid certain logical imperfections which characterize the natural word-languages, they must, of necessity, be still very complicated from the logical point of view owing to the fact that they are conversational languages, and hence still dependent upon the natural languages. (Carnap 1934: 2; emphasis added)

For Carnap, the main deficiency of 'natural word-languages' is that they do not lend themselves to a precise formulation of their own rules of formation and transformation. But, unlike Frege, Carnap is not so much concerned with the perspicuity of *each and every piece of reasoning*; once the rules are precisely formulated, one should proceed in a purely 'formal' way, with no reference to the meaning of the expressions (Carnap 1934: 1). This is a point that could never have been made by Frege, whose concern with content and expressivity remains present throughout (Brandom 1994: 107–8).

Although plausible at first sight – the languages of ordinary life may not be suitable in scientific contexts – there are several conceptual difficulties with this approach. One of them is what Stokhof (2007) refers to as 'the availability assumption': the claim that ordinary languages fail to express a certain content adequately rests crucially on the (contentious) assumption that we have *independent access* to this content, beyond its (defective) expression in ordinary language. Moreover, the expressive argument is compatible with the instrumental, pragmatic view that formal languages are in fact convenient but not essential devices; if the issue is only one of expressivity, suitable regimentations of ordinary language would in principle be sufficiently adequate, even if yielding longer and more convoluted formulations. Now, as will become clear throughout the book, the position to be defended here is that increased precision and perspicuity is a secondary reason why formal languages are such powerful cognitive artefacts; those who emphasize the expressive function of formal languages are arguably merely 'scratching the surface'.

3.2.2 *Depiction*

General concern with expressivity has also led many theorists to formulate the idea that formal languages must strive to be *iconic representations* of whatever it is they represent. Such considerations can be found in Leibniz himself, who suggested that the individual symbols in a *characteristica universalis* should themselves be perspicuously iconic:

We have spoken of the art of complication of the sciences, i.e., of inventive logic . . . But when the tables of categories of our art of complication have been formed, something greater will emerge. For let the first terms, of the combination of which all others consist, be designated by signs; these signs will be a kind of alphabet. It will be convenient for the signs to be as natural as possible – e.g., for one, a point; for numbers, points; for the relations of one entity with another, lines; for the variation of angles and of extremities in lines, kinds of relations. If these are correctly and ingeniously established, this universal writing will be as easy as it is common, and will be capable of being read without any dictionary; at the same time, a fundamental knowledge of all things will be obtained. The whole of such a writing will be made of geometrical figures, as it were, and of a kind of pictures – just as the ancient Egyptians did, and the Chinese do today. Their pictures, however, are not reduced to a fixed alphabet . . . with the result that a tremendous strain on the memory is necessary, which is the contrary of what we propose. (Leibniz 1966: 10–11)

Naturally, the idea that symbols in a formal language should be like hieroglyphs is rather naive, but Leibniz also hints at the idea that the *configurations* of symbols in a language could/should/do have an iconic component. This general idea finds its most famous expression in Wittgenstein's (1963) picture theory of the proposition in the *Tractatus*, which applies not only to formal languages but to meaningful propositions in all languages.[44] Some of the main passages are:

2.2. A picture has logico-pictorial form in common with what it depicts.

3.1431 The essential nature of the propositional sign becomes very clear when we imagine it made up of spatial objects (such as tables, chairs, books) instead of written signs. The mutual spatial position of these things then expresses the sense of the proposition.

3.21 The configuration of objects in a situation corresponds to the configuration of simple signs in the propositional sign.

Wittgenstein does not say much on formal languages specifically, but there is one relevant passage:

3.325 In order to avoid such errors we must make use of a sign-language that excludes them by not using the same sign for different symbols and by not using in a superficially similar way signs that have different modes of signification: that is to say, a sign-language that is governed by logical grammar – by logical syntax. (The conceptual notation of Frege and Russell is such a language, though, it is true, it fails to exclude all mistakes.)

The iconic rationale for formal languages may seem compelling at first sight, but it is also rife with complications. One obvious question to be asked is: if propositions are pictures (in formal languages as elsewhere), what are they pictures *of*? Wittgenstein had a solid story to tell on how propositions depict facts (besides being facts themselves), based on the carefully crafted ontology of the first pages of the *Tractatus*; but, for those of us not prepared to embrace the Tractarian ontology, the issue is a thorny one. The claim that signs and expressions of a formal language depict (and thus refer to) *something* may quite rapidly lead to a reification of entities one may not be prepared to reify, resulting in various forms of realism or Platonism with respect to conceptual entities (and the difficulties these positions entail, as famously argued by Benacerraf). Do they depict 'logical' facts? Do they depict facts in the physical world, but in a 'logical' manner?

---

[44] As with virtually every aspect of the *Tractatus*, the picture theory of the proposition allows for different, conflicting interpretations. Here, I am simply suggesting that, insofar as it emphasizes the pictorial nature of propositions in all languages, it should also apply to *formal* languages.

To be sure, there is room for iconic conceptions of notations not leading to these unpalatable issues, and Peirce's own conception of the role of icons in logical reasoning seems to be essentially motivated by *epistemological* considerations – and, as is well known, these inspired him to develop the most iconic of all formal languages, his Existential Graph.[45]

[The] purpose of the System of Existential Graphs . . . [is] to afford a method (1) as simple as possible (that is to say, with as small a number of arbitrary conventions as possible), for representing propositions (2) as iconically, or diagrammatically and (3) as analytically as possible. (Peirce 1931: § 4.561 n.)

The main claim is that, with diagrams, the Wittgensteinian 'hardness of the logical must' (i.e., the fact that in a deductively valid argument, the conclusion follows of necessity from the premises) can be 'read off' immediately from the disposition of the signs alone, not requiring the interpretive process involved in dealing with purely conventional interpretive systems (Legg 2011). The goal is thus to achieve perspicuity for *epistemological* reasons.

Besides 'traditional' sentential formal languages, we now have a range of diagrammatic formal languages, which some view as more adequate tools for instruction in logic (see in particular Barwise and Etchemendy's 'Hyperproof' system, discussed in Chapter 5). They can be considered as fully fledged formal languages insofar as they are defined by precise rules of formation and transformation, the main difference being that the symbols used and their articulation exploit the two-dimensional features of writing surfaces more extensively (in this sense, even Frege's ideography has a strong diagrammatic component).

At any rate, the iconicity of a well-designed formal language is often one of the motivations offered to depart from other means of expression, and for some people it counts as a criterion in the choice of a formal language to work with (for teaching purposes in particular). Indeed, on the basis of an analysis of a corpus of mathematical equations and logical formulae, Landy and Goldstone (2007a) argued that agents in fact do treat formal notations as diagrams, to a large extent. They identified systematic introduction of spacing which has no theoretical 'formal' role to play and instead seems to have the purpose of aiding agents to visualize the structure of the expression.

---

[45] Legg (2011: 19) argues that iconic representations allow us to 'see' logical necessity much more sharply than symbolic (sentential) representations. 'When one aspect of a diagram *forces* another aspect to be in a certain way, it is this that enables us to "see" necessity. This occurs not by our having epistemic contact with any further "modal object", but by our fully *grasping* the relationships amongst the diagram's different parts.'

Therefore, even if there are philosophical challenges underpinning the conception of formal languages as iconic devices there seems to be empirical evidence corroborating the idea that human agents in fact have a diagrammatic, iconic cognitive relationship with formalisms.

3.2.3 Calculation

Besides expressivity and depiction, a function that is often (but, as I will argue, not often enough!) attributed to formal languages and formalisms in general pertains to the concept of *calculation*, in particular regarding the role that physical signs play in calculative operations. Indeed, we have seen above that for Leibniz all forms of thinking are variations of calculative processes, and moreover that all calculation involves manipulation of symbols. Thus, a well-designed system of notation, a *characteristica universalis*, should first and foremost be an ideal tool for reasoning (viewed as a calculative process), which would allow for the establishment of truths even beyond the domain of mathematics; it would be at once a tool for judgment *and* for discovery – a logic that is not only *demonstrativa* but also *inventiva*.[46] In one of his most famous quotes, Leibniz says:

If this is done [if a *characteristica universalis* is formulated], whenever controversies arise, there will be no more need for arguing among two philosophers than among two mathematicians. For it will suffice to take the pens into the hand and to sit down by the abacus, saying to each other (and if they wish also to a friend called for help): Let us calculate. [*Calculemus!*][47]

Surprisingly, though, Leibniz is one of the few prominent authors to have truly emphasized the calculative benefits of a well-designed formalism. True enough, the general idea of algorithmic methods was one of the building blocks of the 'algebra of logic' tradition (i.e., to use symbolic algebra for the study of logic), but, ironically, Boole's own avowed motivations for the use of symbolic algebra as a tool are essentially *expressive*:

The design of the following treatise is to investigate the fundamental laws of those operations of the mind by which reasoning is performed; to *give expression* to them in the symbolical language of a Calculus, and upon this foundation to establish the science of Logic and construct its method. (Boole 1854: 1; emphasis added) It is designed, in the next place, to *give expression* in this treatise to the fundamental laws of reasoning in the symbolical language of a Calculus. Upon this head it will suffice to say, that those laws are such as to suggest this *mode of expression*, and to give to it a peculiar and exclusive fitness for the ends in view. There is not only a close analogy

---

[46] *On the Art of Combination* (in Leibniz 1966).   [47] As quoted by Lenzen 2004: 1.

between the operations of the mind in general reasoning and its operations in the particular science of Algebra, but there is to a considerable extent an exact agreement in the laws by which the two classes of operations are conducted. (Boole 1854: 4; emphasis added)

Underpinning Boole's approach is essentially a Kantian conception of logic as pertaining to the laws of thought as such, which (according to Boole) find their ideal expression in the language of symbolic algebra, but mainly because algebraic operations themselves (considered independently of their linguistic expressions) mirror very closely the operations of the mind. It is as if the algorithmic advantages would be a bonus, a corollary of the expressive adequacy of the notation.

In truth, however, there is a sense in which expressive and calculative motivations are in tension with each other (see 3.3.4, below). The main rationale behind expressive considerations seems to be the search for *perspicuity*, but perspicuity is often a hindrance when it comes to efficient calculation. As suggested by Krämer, the real strength of calculating methods involving manipulations of signs is that they can be carried out even if the agent does not know why they are effective and correct:

signs can be manipulated without interpretation. This realm separates the knowledge of *how* to solve a problem from the knowledge of *why* this solution functions. (Krämer 2003: 232)

Perspicuity often entails the use of *more* symbols, which in turn yields formalisms that are cumbersome for the purposes of calculation; it is simply cognitively more economical if one can operate with fewer symbols rather than with more. Moreover, and as will be argued extensively throughout the book, the more one 'thinks' about what one is doing when operating with formalisms the more likely it is that external 'intuitions' will interfere in the process, which is something to be avoided in some contexts.

Of course, the use of a given formalism to tackle a particular problem must be based on some degree of *epistemic justification* for the adequacy of the formalism for that application; indeed, when new formalisms are being developed, there is usually a parallel process of search for epistemic justification – not only *that* it works and *how* it works, but also *why* it works.[48] But once a satisfactory degree of epistemic certainty is attained it is no longer necessary to be reminded, each and every time, of why a given formalism is 'trustworthy'.

---

[48] See Heeffer 2007 on the search for epistemic justification in the early development of notations in the abbaco schools treatises.

From this point of view, it also becomes clear why Frege emphasizes perspicuity in his ideography: given that he is involved in a *foundational* project, the goal is precisely to guarantee epistemic justification at every step. Nevertheless, the *Begriffsschrift* notation is so cleverly designed that it also has a strong operative, 'hands-on' component: it is also a tool to reason *with* (Macbeth 2011). It is precisely this performative dimension which was lost when Whitehead and Russell 'translated' the *Begriffsschrift*'s two-dimensional notation into the linear Peano notation. Thus, although this was not Frege's primary goal, the ideography is arguably also a good tool for 'calculation'.[49]

Conversely, while Carnap also seems to have expressivity concerns in mind when motivating the need for especially designed formalisms, his ultimate goal is essentially a calculative one. Once the rules of formation and transformation of a calculus are adequately formulated, reasoning can proceed in a purely 'formal' manner (i.e., with no concern for meaning), which is one of Carnap's main goals.[50] (And notice that Carnap's project goes much beyond the formulation of a language adequate for logic and mathematics alone; it extends all the way into the empirical sciences.) He famously defends the view that languages are (best viewed as) *calculi* (Carnap 1934: § 2), and connects his 'method of syntax' with the anti-metaphysical agenda of the Vienna Circle.[51]

Now, as previously stated, the thesis to be argued for throughout the book is that the powerful cognitive impact of formal languages and formalisms in general is first and foremost of a *calculative* (not expressive) nature. In other words, the present work can be viewed as a forceful defence of the Lullian/Leibnizian idea of a *logica inventiva*: formal languages as cognitive artefacts lend themselves to being manipulated and operated on much more

---

[49] With respect to the two senses of the formal discussed in Chapter 1, it is clear that Frege's ideography is not formal in the sense of de-semantification, but it is formal in the sense of 'computable'. It allows for the fully mechanical application of its rules (i.e., it is a well-defined 'calculus' in this sense), but this mechanical aspect is arguably a means to attain the ultimate goal of epistemic transparency.

[50] Hence also his emphasis on a syntactic approach, rejecting the primacy attributed to judgments by 'prevalent opinion' (Carnap 1934: 1).

[51] It is worth noticing that in the *Tractatus* Wittgenstein also attributes a calculative role to formalisms, but to him this is precisely why they are ultimately superfluous:

> 6.126 One can calculate whether a proposition belongs to logic, by calculating the logical properties of the symbol. And this is what we do when we 'prove' a logical proposition. For, without bothering about sense or meaning, we construct the logical proposition out of others using only rules that deal with signs.

> 6.1262 Proof in logic is merely a mechanical expedient to facilitate the recognition of tautologies in complicated cases.

easily than other representational systems, and this is precisely why they are such powerful tools for the discovery of novel facts and concepts.

3.3 PITFALLS AND LIMITATIONS

It is evident by now that the present work is based on unabashed optimism regarding the strength of formal languages as epistemic, cognitive tools. However, as with any tool, formal languages can also *malfunction*, either because they are not properly designed, or because an otherwise well-designed tool is inadequate for a particular application.[52] It goes without saying that not just any formalization or formalism is appropriate for a given task, and discussions of criteria of adequacy are much needed; yet, sustained analysis of what counts as a 'good' formalization are rare – there are some exceptions (Brun 2003; Baumgartner and Lampert 2008).[53] In Chapter 7, I discuss in more detail the ins and outs of undertaking a formalization (including criteria of adequacy), but in this section I outline some of the limitations and pitfalls of applying formal languages to investigate different classes of phenomena.

3.3.1 *Formal languages are* not *models of ordinary languages*

One appealing view on the epistemic status of formal languages and formalizations more generally is that they are *models*, as understood by philosophers of science (Frigg and Hartmann 2006), i.e., simplified idealizations of some external target phenomena.[54] This view has the advantage of recognizing that, almost inevitably, there will be 'gaps' between models and phenomena (which is an improvement over the view that the formalism perfectly matches the phenomena, or that it captures something really *in* the phenomena, e.g., the 'logical form' of sentences), but it requires a serious reflection on what it is that formal languages and formalisms are models *of*.

According to a popular view, formal languages are models of *vernacular/ ordinary languages*:[55]

A formal language is a *mathematical model* of a natural language (of mathematics), in roughly the same sense as, say, a Turing-machine is a model of calculation, a

---

[52] Høyrup (2007) discusses a few historical cases of artificial languages of mathematics arguably having been 'excessively generous' and 'delusive'.

[53] However, both focus almost exclusively on formalization as a process going from portions of ordinary language to formal languages, a view to be criticized shortly.

[54] I use the blanket term 'target phenomenon' to refer to whatever it is that a formalism seeks to capture or characterize.

[55] Most often referred to as 'natural languages', a terminology criticized in Chapter 2.

collection of point masses is a model of a system of physical objects, and the Bohr construction is a model of an atom. In other words, a formal language displays certain features of natural languages, or idealizations thereof, while ignoring or simplifying other features. (Shapiro 1998: 137)

There is almost always a gap between a model and what it is a model of. (Shapiro 2006: 50)

On the Shapiro/Cook view, a logic is not meant to be a perfect representation of the linguistic phenomenon being studied. Instead, a logic is a model of that linguistic phenomenon, and as a result, all of the advantages and limitations that are present in modeling elsewhere (such as in the empirical sciences) should be expected to reappear in the study of logic. (Cook 2010: 500)

Prima facie, this sounds like a reasonable position. However, it does not accord with the historical development of formal languages; as argued in section 3.1, modern formal languages did not emerge from the *logical* tradition of regimentation and schemata, but rather from the *mathematical* tradition of algebraic notations. Furthermore, this position either disregards or does not sufficiently emphasize the operative, 'hands-on' import of formal languages and formalisms, viewing them essentially as expressive devices.

More importantly, it is not at all obvious which portions or fragments of ordinary languages formal languages are models *of*, given some striking discrepancies (as observed empirically by linguistic and psychological studies). In effect, virtually all traditional logical constants are typically given a semantics that deviates considerably from the usual meaning of their 'counterparts' in ordinary language: the conjunction is viewed as commutative ($A \wedge B$ is equivalent to $B \wedge A$), whereas ordinary language 'and' is notoriously order-sensitive; the disjunction is not exclusive ($A \vee B$ is compatible with both A and B being true), whereas ordinary language 'or' is typically exclusive (Chevallier, et al. 2008); implication does not allow for exceptions, whereas recent research (Stenning and van Lambalgen 2008) suggests that people seem to attribute defeasible interpretations to 'if . . . then' formulations; the universal quantifier ranges over the total universe of discourse, whereas ordinary speakers usually understand 'every' or 'all' as implicitly ranging over a particular domain of discourse (Counihan 2008a).

And yet, generations and generations of students are taught how to 'translate' ordinary-language statements into first-order logic in 'intro to logic' classes, as if this was the ultimate application of the formalism; no wonder that this is such a resilient (although erroneous) ideology.[56]

---

[56] 'When introducing logic, I teach the usual skill of "translating" natural language into the formal language of first-order logic, and in my didactical prose, I unthinkingly copy what generations of

Instead, I suggest that it is best to view the traditional logical constants as in fact denoting some very basic *operations*, which taken together yield a powerful formalism (both for calculation and for expression); in other words, something along the lines of Boole's conception of logic, minus his psychologistic grounding for 'the laws of thought'.

It may be argued that formal languages are models of ordinary languages in specific contexts, *mathematics* in particular, as Shapiro's en passant, parenthetical qualification in the passage above suggests. This is a more plausible position, in the spirit of the Fregean expressive project of designing a language to formulate perspicuous, 'gapless' versions of mathematical proofs. But this would restrict the scope of application of formal languages to mathematical contexts, and while this was indeed Frege's original intention, the development of logic since has gone much beyond that.

Thus, I submit that the view according to which formal languages are models of linguistic phenomena is misguided. The assumption seems to be that, to describe a given target phenomenon, a formal language or formalism makes a detour via ordinary language discourse about the target phenomenon in question:

$$\text{Formal language} \longrightarrow \text{Ordinary language discourse about phenomenon } \aleph \longrightarrow \text{Phenomenon } \aleph$$

Instead, I claim that a more fruitful way to think about formal languages as models is:

$$\text{Formal language} \longrightarrow \text{Phenomenon } \aleph$$

In this sense, a formal language will be much like any mathematical formalism, a 'language' in itself which can characterize directly the target phenomenon without the mediation of ordinary languages.[57] To illustrate, consider the mathematical formalism developed by Maxwell within his theory of electromagnetism (to be discussed in Chapter 6); it was not

---

colleagues have said about it. But recently . . . it suddenly occurred to me that I no longer believe in these formulations. Or maybe worse, I no longer understand what we are claiming about this skill, and hence what we are teaching' (van Benthem 2011)

[57] This conclusion may appear to be at odds with my endorsement (end of section 3.2.1) of Stokhof's (2007) critique of the 'availability hypothesis'. As will become clear in Chapter 7, the view defended here is that the ordinary language formulation of a theory or idea does provide *epistemic access* to its conceptual content, but is not itself the object of formalization.

intended to regiment ordinary discourse about electromagnetic phenomena, but rather to describe directly and in *its own terms* the phenomena in question.

The same holds, *mutatis mutandis*, when the phenomena in question are not of a physical but rather of a conceptual nature. For example, in my own work of formalizing medieval logical theories (Dutilh Novaes 2007), the procedure was never to engage specifically with the textual formulation of these theories (in a 'translation enterprise', as it were), but rather to absorb for as much as possible the conceptual structure of these theories through the text, and then to formulate it in a given formalism – what I have described (Dutilh Novaes 2007: chap. 4) as 'conceptual translation'. My suggestion here is thus that to view formal languages as models of ordinary languages is not only misguided but potentially pernicious. 'Modelling' ordinary languages is the wrong use for these otherwise versatile artefacts.[58]

3.3.2 System imprisonment

It was mentioned above that the shift towards viewing formal languages as something *about* which one can prove theorems, i.e., as mathematical objects in themselves, greatly contributed to the progress of the meta-theoretical programme consolidated in the 1920s (Awodey and Reck 2002). It was also suggested that, while proving meta-theorems about formalisms was a much-needed step for the establishment of their epistemic justification, the emphasis on metalogic also had the effect of deflating the relevance of logical formalisms as tools *with* which to prove theorems, in the spirit of Frege's *Begriffsschrift*. But there were other significant consequences of the objectification of logical systems, in particular what van Benthem (2011) describes as 'system imprisonment'. Here are some relevant passages:

> But how good is the model of natural language provided by first-order logic? There is always a danger of substituting a model for the original reality, because of the former's neatness and simplicity. I have written several papers over the years pointing at the insidious attractions and mind-forming habits of logical systems. Let me just mention one. The standard emphasis in formal logical systems is 'bottom up'. We need to design a fully specified vocabulary and set of construction rules, and then produce complete constructions of formulas, their evaluation, and inferential behavior. This feature makes for explicitness and rigor, but it also leads to *system imprisonment*. The notions that we define are relative to formal systems. This is one of the reasons why outsiders have so much difficulty grasping logical results: there is usually some parameter relativizing the statement to some formal

---

[58] Except, of course, when the very phenomena to be modelled (the target phenomena) *are* ordinary languages themselves, as in, for example, some branches of linguistics or formal semantics.

system, whether first-order logic or some other system. But mathematicians want results about 'arithmetic', not about the first-order Peano system for arithmetic, and linguists want results about 'language', not about formal systems that model language. (van Benthem 2011: 3)

I am worried by what I call the 'system imprisonment' of modern logic. It clutters up the philosophy of logic and mathematics, replacing real issues by system-generated ones, and it isolates us from the surrounding world. I do think that formal languages and formal systems are important, and at some extreme level, they are also useful, e.g., in using computers for theorem proving or natural language processing. But I think there is a whole further area that we need to understand, viz. the interaction between formal systems and natural practice.[59]

There are many important points made in these passages. First, recall that in the early days of modern axiomatics (end of the nineteenth century), proofs were given with a tacitly assumed and (presumed to be) unique concept of deductive consequence, which bestowed these proofs with a certain aura of absoluteness. One essential development in the aftermath of *Principia Mathematica* was the explicit, rigorous formulation of deductive systems, thus giving birth to the modern notion of a formal system. This had two opposite epistemic consequences: on the one hand, because now desirable meta-properties of the systems could be proved (consistency, completeness), the reliability of proofs within them increased; but, on the other hand, the validity of these proofs was now relative to the formal system in question, and there could at least in theory be several equally 'good' formal systems. There was no longer an absolute sense of validity or proof to be relied on.[60] Carnap (1934) represents the self-conscious pinnacle of these developments, with his suggestion that there is no absolute perspective for proofs, only proofs relative to a language/calculus.

As observed by van Benthem, 'mathematicians want results about "arithmetic", not about the first-order Peano system for arithmetic'; in other words, they want results about the target phenomenon, the 'real thing', not about the formalism used to describe it. And yet, logicians working with formal systems are first and foremost proving theorems *about the formalism*, which say something about the target phenomenon only insofar as the fitting between formalism and phenomenon is good enough. Thus, the

---

[59] Extract from an interview which appeared in Chinese in *Philosophical Trend* 1 (2008) (English version kindly provided by J. van Benthem).

[60] This is of course a variant of classical foundationalist problems. But notice that even Hilbert's project of proving the consistency of arithmetic within arithmetic relies on a form of dubious circularity; indeed, if arithmetic was inconsistent, it would be unsound (assuming the classical connection between inconsistency and trivialization), and thus it might be able to prove its own consistency even if it was in fact inconsistent.

upshot of working with rigorously formulated formal systems is a form of *system imprisonment*.

In a sense, system imprisonment is the trade-off for the benefits of working with formal systems which can be studied as mathematical objects themselves; arguably, it is just something we have to live with. But another, potentially more pernicious consequence of working with formal systems is what van Benthem describes as 'system-generated issues'. The idea is that, while working with a given formal system, a particular technical issue may arise within the system, but which is then viewed as a 'real issue' pertaining to the target phenomenon, rather than a property of the formal system itself. Van Benthem's favourite example of a system-generated issue is the issue of sharply demarcating first- from second-order logic; according to him, this issue is purely internal to the logical systems, having no 'external' counter-parts, and yet it is still seen as a crucial, deep philosophical question.[61]

My own favourite example of a system-generated problem pertains to the formulation of de re modal statements in classical quantified modal logic. As is well known, while in de dicto propositions the modal operator ranges over a complete proposition (a closed formula with no free variables), in a de re modal proposition the modal operator is introduced between the quantifier and the variable it binds. A de dicto proposition such as 'Possibly there is something which is greater than seven' is rendered as:

$$\Diamond \, \exists x \, (x < 7)$$

A de re proposition such as 'There is something which is possibly greater than seven' is in turn rendered as:

$$\exists x \Diamond (x < 7)$$

From the start, there is the technical problem of formulating a composi-tional, recursive semantics for a language where an operator ranges over an open sub-formula, i.e., with a free variable (to be sure, there are ways around this technical issue). More important, however, is the fact that this specific rendition of de re modal propositions was and still is seen as entailing all kinds of undesirable metaphysical commitments, in particular unpalatable forms of essentialism.

[I]f to a referentially opaque context of a variable we apply a quantifier, with the intention that it govern that variable from outside the referentially opaque context, then what we commonly end up with is unintended sense or nonsense of the type

---

[61] Personal communication.

(26)–(31) [examples omitted]. In a word, we cannot in general properly quantify into referentially opaque contexts. (Quine 1953: 148)

In other words, the claim is that de re modality requires that the modal operator be sandwiched between the quantifier and the occurrence(s) of the variable it binds, but in practice this procedure severely disrupts the quantificational effect. Clearly, though, the issue only arises in virtue of a particular artefact of classical predicate logic, namely the fact that quantification is treated by means of *variables*. As much as one may think that this is the only way to treat the concept of quantification, this is simply not true. The predominance of predicate logic is such that we cannot (easily) conceive of alternative approaches to quantification, where the problem of 'free variables' might not arise: we are 'trapped' inside the system. It thus becomes difficult to appreciate that the issue of 'referentially opaque contexts' is purely internal (system-generated) rather than inherent to quantification and modality as such.[62]

In conclusion, I submit that failure to separate features that are artefacts of the formalism from issues that pertain to the phenomenon being studied as such is an important drawback of using formal methods to investigate external 'realities'. A certain degree of system imprisonment is inevitable, but we should remain alert for the risk of attributing external reality to specificities of the formalism.[63]

3.3.3 *Intrinsic limitations: deductive incompleteness and non-categoricity*

Once more, let us go back to the development of formal axiomatics in the late nineteenth and early twentieth centuries. As discussed by Awodey and Reck (2002: 5), the main avowed goal of the different axiomatizations of portions of mathematics proposed at the time was 'to treat the objects of mathematical investigation more abstractly, and then to characterize them

---

[62] For an alternative approach to modality and quantification, see Dutilh Novaes 2004. There, I also argue that, contrary to common belief, it is not the case that 'traditionally a modality de dicto was seen as an attribution of necessary (or possible) truth to a proposition (*dictum*), and a modality de re was seen as an attribution of necessary (or possible) property to an entity (*res*). The traditional distinction corresponds to the formal one' (Gamut 1991: 47). In fact, this so-called 'traditional' way of thinking about de re modality was essentially induced by the rendition of such propositions in quantified modal logic, first proposed by von Wright (1951).

[63] To be sure, *some* features of the formalism must have counterparts in the phenomenon being modelled, otherwise the formalism would simply not be an adequate model. Shapiro (1998) introduced the convenient distinction between 'representors' and 'artefacts' within a formalism; Dutilh Novaes 2011d provides a discussion of this terminology, and the difficulty seems to consist precisely in distinguishing artefacts from representors in specific cases.

completely'. But what 'completely' means here must be spelled out, as at the time there were different and for the most part vaguely (if at all) formulated notions of completeness floating around. Presently, there are also several different notions of completeness on the market, but two familiar ways of formulating the completeness of a given theory T with respect to mathematical object(s) MO it seeks to characterize are:[64]

- For every formula φ in the relevant domain of discourse, T either proves φ or it proves its contradictory.
- For every formula φ, if what φ says is the case in MO, then T proves φ.

In both cases, what is in question is a theory's *deductive* strength, i.e., its ability to prove everything that is the case with respect to the underlying mathematical object/structure. Indeed, here I will be concerned primarily with *deductive* completeness in these senses.

From the very start, one of the main avowed goals of axiomatizations and formal systems in general is to achieve deductive completeness, i.e., to be able to prove all that is the case with respect to the target phenomenon.[65] But, besides characterizing a mathematical object/structure completely (in this sense), another plausible desideratum for an axiomatization/formal system is that it characterizes *nothing but* the intended target phenomenon. The property of a theory of characterizing only its intended target phenomenon is known as *categoricity*:

- A theory T is called categorical (relative to a given semantics) if for all models M, N of T, there exists an isomorphism between M and N.

In other words, if M is a model of T, then any other model of T is in fact identical to M, in an appropriate sense of 'identical' (typically, in terms of isomorphism). Now, completeness and categoricity are typically both seen as central goals in formulations of a formal system/theory, but, as is well known, under numerous circumstances they can simply not be obtained.

Historically, the realization that categoricity would not follow necessarily from a well-crafted axiomatization precedes a similar realization with respect to deductive completeness. Indeed, Dedekind himself was aware of the fact that his axiomatization for arithmetic without the final second-order axiom was not categorical in that it also characterized structures that were very different from the intended series of the natural numbers (Dedekind 1890).

---

[64] Awodey and Reck (2002: 2–4) list other formulations, in particular notions of *semantic* completeness, which will not concern us here. See also Hintikka 1989: sect. II.
[65] In the early days of axiomatics, the target phenomena were predominantly mathematical objects and structures, such as the series of natural numbers and the arithmetic operations pertaining to it.

In contrast, that deductive completeness could not be taken for granted was a hypothesis entertained for the first time only much later; a farsighted comment by the American postulate theorist Oswald Veblen in the 1900s is perhaps the first formulation of this possibility:

But if [a proposition] is a consequence of the axioms, can it be derived from them by a syllogistic process? Perhaps not. (Veblen 1906: 28)

As is well known, however, it was only with Gödel's incompleteness results in the early 1930s that it became apparent that the (deductive) completeness of a formal system with respect to the phenomenon it seeks to characterize is by no means to be taken for granted. We now know that any formal system containing arithmetic and having sufficiently strong expressive properties (diagonalization in particular) cannot, by definition, be complete. But incompleteness (in different guises) is actually a more general phenomenon, pertaining not only to systems containing arithmetic; second-order logic, for example, is deductively incomplete with respect to its 'standard' semantics, although it *is* complete with respect to Henkin semantics, as noted by Shapiro.[66]

Incompleteness and non-categoricity can been seen as duals of each other in the following way: incompleteness pertains to the failure to capture (deductively) all the 'truths' that one seeks to capture with a formal system or theory, while non-categoricity pertains to the failure to characterize the target phenomenon *uniquely*. Speaking metaphorically, in the former case, not enough fish are caught by the net of the formal system; in the latter case, too many fish are caught. Moreover, there are also interesting technical connections between the two concepts: if a formal system FS is incomplete with respect to the (mathematical) objects it seeks to characterize, then there is a formula ϕ that is neither proved nor disproved by FS. And, thus, FS is compatible both with objects/structures where ϕ is the case and with objects/structures where ϕ is not the case, yielding non-categoricity.

Incompleteness and non-categoricity are thus intrinsic limitations in the project of characterizing a given class of objects or phenomena completely and uniquely by means of an axiomatization/formal system. To be sure, there are cases of systems that are complete and categorical (relative to the desired structures/phenomena), but when these limitations do arise, it is not due to an imperfect fitting between phenomena and formalism which could, in principle, be remedied with the formulation of more precise

---

[66] 'It is not a matter of having left out one or two invalid deductions. *No* effective deductive system is sound and complete for second-order validity – assuming standard semantics' (Shapiro 2006: 654).

rules and axioms (within the same language); rather, it arises from limitations intrinsic to the formalization enterprise itself.

Non-categoricity pertains to the degree of expressive power of a theory, while deductive incompleteness pertains to its deductive strength. For example, as is well known, it is impossible to formulate a categorical axiomatization of arithmetic in a first-order language; such languages are simply not sufficiently expressive. To obtain a categorical axiomatization of arithmetic, one must resort to second-order expressive devices, which then bring along a series of unwanted features (starting with the fact that the second-order consequence relation does not admit a complete proof procedure). Deductive incompleteness, in turn, concerns the very limits of human knowledge by means of deductive methods; the rationalist ideal of attaining all truths within a given domain (departing from the right premises/axioms) by means of deductive chains simply cannot be realized in a few crucial cases. Formal methods may go very far, but they also have their own intrinsic limitations.

Despite this, one must be wary of excessively pessimistic conclusions concerning the power of formalization (and of human cognition in general) drawn on the basis of the phenomena of non-categoricity and incompleteness. In a number of cases, completeness and categoricity do obtain (although often enough, one comes at the expense of the other – see next section), and at any rate soundness alone already guarantees the (relative) utility of a given formal system. But, more importantly, it must not be forgotten that the groundbreaking results on incompleteness and non-categoricity were obtained *precisely* thanks to the power of formalizations and formal systems: this methodology has the remarkable capacity of investigating its own limits.[67]

3.3.4 *Intrinsic limitations: tractability v. expressiveness*

When discussing the expressive function of formal languages in section 3.2.1, I had in mind a non-technical sense of expressiveness, namely the idea that formal languages can express/describe concepts and ideas that have an independent, prior existence. There is, however, a technical sense of expressiveness, current in logic and computer science, which pertains to the expressive power of a language as a mathematical concept: the range of properties of mathematical structures that a language can (or cannot)

---

[67] In fact, in my opinion, the 'foundations of mathematics' project is at its best when it does exactly this: explore and investigate its own limits in a technical, rigorous way.

discriminate. In the previous section, when discussing the issue of (non-) categoricity, the technical notion of expressiveness was already alluded to: first-order languages are not sufficiently expressive to rule out unintended models of arithmetic, thus yielding non-categoricity. It is only with the augmented expressive power of second-order languages that categoricity for arithmetic can be obtained. However, once a second-order logic underpins a given theory, there can be no complete proof procedure for it, as second-order logic has adverse deductive properties. Thus, in several instances, the deductive price to be paid for categoricity is very high, perhaps too high.

The (non-)categoricity of theories of arithmetic is but a special case of a more general phenomenon, widely discussed by computer scientists: the orthogonality of the desiderata of expressiveness and tractability. As a general rule, the more expressive a language (or programme) is, i.e., the more terms it contains, the less tractable it is, i.e., the less favourable are its computational properties. The notion of tractability currently entertained by computer scientists encompasses the idea that a procedure must be computably effective *within a reasonable timeframe*, usually in terms of polynomial-time computability; so it is not sufficient that it be implementable, it must also be implementable with a certain speed. Indeed, a classical theme in computer science and artificial intelligence is the *trade-off* between tractability and expressiveness (Levesque and Brachman 1987): you just cannot have your cake and eat it. Ideally, you must find the point that optimizes the combination of expressive power and tractability for a given task.

The orthogonality of expressiveness and deducibility[68] has also been discussed by philosophers. Hintikka (1989) distinguishes two functions of logic and logical apparatuses in mathematics: the *descriptive* use and the *deductive* use. The descriptive use is what underpinned the pioneer work on the foundations of mathematics of Dedekind, Peano, Hilbert, etc.: the formulation of sets of axioms which would characterize (describe) a unique mathematical structure (or a unique class of structures) *completely* (Awodey and Reck 2002; see also section 3.1.2, above). In these early days, only the descriptive use of logic was being deployed, except for Frege who already assigned a clear *deductive* use to logic in mathematics. With his ideography, Frege sought to capture the *chains of reasoning* of mathematicians when proving theorems rather than simply describing mathematical structures. The deductive perspective only became widely adopted after *Principia*

---

[68] Tractability is a more restricted concept than deducibility, as it involves the time parameter; but the general phenomenon emerges even when the time it takes to compute is disregarded.

Mathematica, i.e., after a formalized approach not only to axioms but also to rules of inference became available.

The orthogonality of descriptive v. deductive desiderata can again be illustrated by the case of formal theories of arithmetic. As we have seen, first-order theories of arithmetic are non-categorical in that they do not have sufficient expressive power to exclude unintended models; but moving from first- to second-order theories entails a significant loss of deductive power. As is well known, first-order logic is complete and has (generally speaking) favourable computational properties;[69] second-order logic, by contrast, is a computational disaster. So, in the case of theories for arithmetic (as elsewhere), both desiderata cannot be simultaneously satisfied: one goes at the expense of the other. Thus, one conclusion to be drawn is that the choice of formalism will also depend on the goal of the formalization: if descriptive, then it is advantageous to choose a formalism with a high degree of expressive power; if deductive, then one is better off with a less expressive but more tractable formalism.[70]

In the context of the present analysis, the technical impossibility of combining expressive power and tractability also illustrates the orthogonality of the expressive and depictive functions attributed to formal languages, on the one hand, and the calculative function, on the other hand (section 3.2). Even though the informal concept of expressiveness developed in section 3.2.1 is not exactly equivalent to the technical concept of expressive power, the general idea is quite similar. Throughout the book, I develop the idea that expressively strong formalisms (in the informal sense) may interfere with the reasoning process, and thus hinder calculation, on account of facilitating the intrusion of external, irrelevant beliefs. A different but related point is the observation that expressiveness and tractability do not go hand in hand, as extensively investigated in the computer science and artificial intelligence literature.

3.4 CONCLUSION

The aim of this chapter was to present a brief overview of the historical developments leading to the emergence of formal languages and their uses

---

[69] But it is not decidable; for example, there is no procedure to determine, for any formula, whether there can be a proof for it or not.

[70] Historically, it would seem that the descriptive use of formalisms was predominant in most of the twentieth century, in particular with the establishment of model theory and the consolidation of the idea of using logical tools to describe mathematical structures. But, in recent years, with the growing influence of computer science-oriented approaches in logic, the deductive use is regaining strength.

in logic. I have argued that, rather than in the logical tradition based on regimentation and schemata, formal languages have their origins in the mathematical tradition of algebraic notation (section 3.1). I have also surveyed the traditional rationales attributed to using formal languages in logic, again contrasting expressive v. calculative motivations (section 3.2). These historical observations lend further support to my claim that the real cognitive impact of formal languages pertains to the realm of calculability/computability rather than to the realm of expressivity, which will be argued for throughout the book. Moreover, one of the upshots of this position is the view that formal languages are not to be seen as models of ordinary languages (section 3.3.1). But I have also explored some of the risks and limitations of this investigative methodology, highlighting the concepts of system imprisonment, system-generated problems (section 3.3.2), incompleteness and non-categoricity (section 3.3.3), and the impossibility of maximizing both expressiveness and tractability (section 3.3.4).

With this historical and philosophical background in place, we are now ready to embark on the truly empirically informed part of the book.

II

CHAPTER 4

How we do reason – and the need for counterbalance in science

In order to assess the cognitive impact of reasoning *with* formal languages, which is the main topic of the present investigation, we must first come to grips with how human agents typically reason *without* formal languages. Of course, this is much easier said than done: even after decades of intensive research, psychologists and cognitive scientists still have not been able to produce a definitive account of the mechanisms underlying reasoning in human beings. There is still substantial disagreement even on the outlines of the story, and many open questions are still in the air. Nevertheless, at least some findings seem sufficiently robust so as to allow for a few (still tentative) hypotheses on some of the mechanisms underlying human reasoning.[1] In this chapter, I review what I consider to be the main findings in the literature, and attempt to formulate a coherent picture of human reasoning.

It is customary in psychology and cognitive science to differentiate between different levels of analysis. The *locus classicus* for this is Marr (1982), who distinguished between the computational, the algorithmic, and the implementation levels in cognition. There have been different proposals on how to define the different levels of analysis and what terms to use (Stanovich 2008: 414–16), but generally what Marr termed the 'computational level' and others refer to as the 'intentional level' pertains to the *goals* of the cognitive processes in general and the function(s) that the system in question is serving. The algorithmic level concerns the actual details of the processes in question, how they actually unfold; finally, the implementation level concerns the instantiations of the processes described

---

I have greatly benefited from detailed comments by Shira Elqayam and Keith Stenning on an earlier draft of this chapter.

[1] As will become clear throughout the chapter, I am highly indebted to the work of Keith Stenning (both his solo work and his collaboration with Michiel van Lambalgen) in my views concerning human reasoning.

by the algorithmic level in physical systems. Let me make it clear from the start that, in this chapter, we will be concerned predominantly with the algorithmic level of analysis, unless otherwise stated. (In Chapter 5, the 'implementation level' will also occupy a central position.)

4.1 'DISPROVING PIAGET'

In the early days of research on reasoning within experimental psychology (in the first half of the twentieth century), the fact that participants' performance often deviated from the normative responses (as defined by the canons of logic traditionally construed) had already been noticed (Wilkins 1928; Morgan and Morton 1944). Nevertheless, a milestone in this research tradition is the work of Piaget and collaborators, most notably Inhelder.[2] Some relevant aspects of Piaget's conception of human cognition and rationality are described in the following passage:

> How do people reason? The view that I learned at my mother's knee was that they rely on logic.[3] ... Jean Piaget, and his colleagues argued that the construction of a formal logic in the mind was the last great step in children's intellectual development, and that it occurred at about the age of twelve (see, e.g., Inhelder and Piaget, 1958). (Johnson-Laird 2008: 206; cf. Inhelder and Piaget 1958)

In other words, the predominant idea was that (mature, adult) reasoning is rule-based, and that the 'abstract' rules in question were most likely the rules of 'logic' as traditionally construed (syllogistic, classical logic). Notice, however, that Piaget did not claim that we were *born* with these rules in our heads, ready to use as it were; the rule-based system of reasoning had to mature, and in principle education might play a fundamental role (in the sense that the system might not mature if not exposed to adequate training). Indeed, the aspect of Piaget's notion of 'formal operations' which has attracted most attention is the idea of reasoning proceeding by schematic substitution with different content (the formal as schematic),[4] but Piaget's

---

[2] In fact, it seems that psychologists are still struggling to set themselves free from the Piagetian framework. Their results typically go against Piaget's theses, but the conceptual framework they operate with is still very much that of Piaget.

[3] Johnson-Laird mentions Boole and Mill as precursors of this view, but the real (albeit indirect) source of the idea that logic concerns the 'laws of thought' and 'correct reasoning' (as opposed to the art of *argumentation*) is Kant. Ultimately, and unbeknownst to them, the whole psychology of reasoning tradition seems to rest on Kantian assumptions which are uncritically accepted (Dutilh Novaes 2011d). But the normative component of Kant's conception of logic was somehow lost along the way; as the passage by Johnson-Laird indicates, the idea is that we not only *should* but in fact *do* reason relying on logic.

[4] See Dutilh Novaes 2012 for the formal as schematic.

doctrines are in fact significantly more sophisticated than they are often portrayed to be by later researchers who claim that Piaget's ideas have been entirely disproved.[5]

This is not the place for an extensive analysis of Piaget's ideas, but it is important to add a word of caution concerning the discrepancies between Piaget's actual doctrines (which are, admittedly, often obscurely formulated) and the 'Piaget' that later researchers argued against. What matters for our purposes is that Piaget became associated with the idea of an almost inexorable developmental path towards formal, abstract reasoning as the pinnacle of a child's cognitive development, which in turn was interpreted as entailing that people generally rely on logic for reasoning.[6] But Piaget himself emphasized the role of education in the process of acquisition of (classical) logical competence, which is a key component of the conception of deductive competence underlying the present investigation.

Be that as it may, the picture of human reasoning as proceeding by means of instantiations of rules and schematic substitution with different content soon (i.e., starting in the late sixties) began to be challenged by a series of experimental findings suggesting that humans do not really reason on the basis of logical rules, or in any case not the specific logical rules pertaining to a traditional conception of logic. Of course, the question remains whether human reasoning is not rule-based *at all*, or whether it *is* ruled-based, but just by different, still to be discovered, rules (again, on an algorithmic level of analysis). In what follows, I present an overview of some of the results that essentially discredited the 'Piagetian' picture of human rationality as tied to a particular conception of logic.

4.1.1 *Ordinary reasoning does not follow the rules of (classical) logic*

The first severe blow to the Piagetian paradigm was inflicted by the results of Wason's famous 'selection task' experiment (Wason 1966).[7] Although this is probably the best-known and most discussed experiment in psychology of reasoning, it is perhaps still useful to summarize the experimental set-up and the results here.

---

[5] I owe the point concerning the complexity of Piaget's doctrines to Keith Stenning.
[6] Notice also that the conception of logic underlying this thesis, which is in fact essentially the one that psychologists still have in mind when they speak of 'logic', is very narrow; it emerged basically from old logic textbooks. Recent developments within logic as a discipline go well beyond this narrow conception.
[7] In Johnson-Laird's terms, 'The event that woke me from my dogmatic slumbers was the late Peter Wason's discovery of the effects of content on his "selection" task' (2008: 206).

Participants are shown four cards, for example [A], [B], [4], and [7], and told that each card has a number on one side and a letter on the other side. They are then asked to turn all and only the cards they must turn in order to verify the following conditional: 'If a card has an even number on one side, then it has a vowel on the other side'. Originally, the assumption was that participants would (should) interpret the 'if ... then ...' clause as having the logical properties of the material conditional; in that case, the conditional statement would be falsified by cards with an even number on one side and a consonant on the other side. From that point of view, naturally, one should turn all cards which show either an even number or a consonant to see what is on the other side. Another way of saying the same is that this conditional is, according to classical logic, equivalent to its contrapositive, namely 'If a card has a consonant on one side, then it has an odd number on the other side'. Thus, it is clear that cards showing consonants should also be turned (that is, again, if the conditional is interpreted as a material implication). Indeed, the response predicted by the (Piagetian) assumption that participants would be using classical logic (competently) to solve this task is that they would turn [4] and [B].

But this is not what the great majority of participants do: in fact, less than 10 per cent of the participants turned [4] and [B]. A significant number of participants turned only [4], many others turned [4] and [A].[8] Clearly, they were not using (classical) logic to solve the task, either because they did not interpret the conditional as a material implication, or simply because they were not following the rules that determine the meaning of the material implication according to classical logic. Hence, these results were seen as incompatible with the view that adults reason on the basis of 'formal' rules (formal in the sense of schematic, which can then be instantiated by different occurrences of the 'gaps'). Importantly, participants had not received specific training in logic, but they were undergraduates with many years of formal education behind them. Again in the words of Johnson-Laird:

> The failure was embarrassing to Piagetians, because Piaget had argued that once children attain the level of 'formal operations' – the level corresponding to the acquisition of logic – they would check the truth of conditionals of the form, *If A then B*, by searching for counterexamples of the form: *A and not-B*. (Johnson-Laird 2008: 206–7)

---

[8] Typical results for the original version of the Wason selection task (adapted to the formulation used here): 35% turn [4]; 45% turn [4] and [A]; 5% turn [4] and [B] (Stenning and van Lambalgen 2008: 49).

Participants do realize that they must turn the cards that fit the description in the antecedent (virtually all of them turn [4]), but they fail to realize that they must also turn the cards that *do not* fit the description in the *consequent*. (The results have been replicated several times.) Why is that? Logic alone cannot explain this preference for the positive version of the consequent.[9] Why is it so hard to see that one must turn [B] to verify the conditional (that is, *if* it is a material implication)? Wason and collaborators then embarked on a series of similar experiments, manipulating different elements of the task in attempts to elicit a higher proportion of 'normative responses' (Johnson-Laird 2008: 207), but in first instance with no success.

As if this was not a sufficient blow to the Piagetian conception of reasoning, the experiment where participants finally performed much closer to the 'normative response' was the one where, instead of using 'abstract' material (letters and numbers), cards with actual 'content' were used (Wason and Shapiro 1971). The conditional used was 'Every time I go to Manchester I travel by train', and the cards used were [Manchester], [Leeds], [train], [car]. Unlike in the 'abstract' version of the task, participants were much more likely to realize they must also turn [car] to see if it does or does not have [Manchester] on the other side (10 out of 16 gave the 'correct' answer). These results have been replicated several times, with different kinds of 'contents', and it is clear that the use of contentual material often (though not always) has a significant facilitating effect. What exactly is the nature of this facilitating effect is still an open question, and different views have been proposed, as we will see shortly.

The 'Piagetian' view of human reasoning predicts that there should not be a difference in reasoning performance with 'abstract' or 'contentual' material; if anything at all, it might even predict that performance would be better with 'abstract' material, as this would avoid the additional step of actually interpreting the formal rules presumably guiding reasoning with a specific content. In the case of 'abstract' material, the rules would already be in their 'pure form', so the reasoner would not have to perform the additional step of looking for the appropriate form by separating form from content.[10] And yet, these are not in any way the results that emerge from these experiments.

---

[9] This seems at least to some extent to be related to what is referred to as 'matching bias', to be discussed in Chapter 6.

[10] Notice how much the 'form v. content' ideology permeates the Piagetian doctrine; it is still largely presupposed by psychologists, even though within philosophy its purported self-evidence has been contested.

These results convinced many researchers that formal logic was an entirely inadequate tool to investigate human reasoning: presumably, humans do not reason on the basis of 'formal', abstract rules, otherwise content would not have a facilitating effect. Of course, and as has been argued by Stenning and van Lambalgen (2008) and by Rips (1994), this rejection is based on a very narrow conception of logic. In any case, it became clear that the view of human reasoning as being governed by the abstract rules of (classical) logic which are then interpreted in specific cases (essentially the procedure of instantiating logical schemata in particular cases/domains) did not seem to offer an accurate picture of the subject-matter. For our purposes, it is crucial to keep in mind that (untrained) human agents typically reason 'better' when they reason with content rather than with schematic letters and symbols. As I will argue later on, reasoning with symbols is a technique which must be specifically trained for in order to be mastered; it does not develop spontaneously.

Nevertheless, the idea of reasoning being rule-based continued to be defended well into the 1990s (I am not aware of current proponents of it), in particular by Rips (1994) and by Braine and O'Brien (1998). Rips (2008: 167)[11] correctly notes that psychologists had remained oblivious to the latest developments in logic introducing a wide variety of logical systems, and that they still assumed that there was only one standard of correct deductive reasoning. So, if experiments showed that participants did not reason according to the traditional rules of logical inference, it still did not mean that they did not reason according to *any rules at all*. Rips (1994) proposed an account of human reasoning as largely based on something like a natural deduction proof-system, and claimed that it was successful in accounting for many of the experimental data. The 'enemy' at that point was the mental models theory of human reasoning, defended by Johnson-Laird (2008) and collaborators (see also below).

Presently, the consensus seems to be that the thesis that there is a 'proof system' underlying human reasoning, largely following the structure of familiar natural deduction systems (even if with rules other than those of classical logic), is untenable, both for conceptual and for empirical reasons. Rips (2002) himself seems not to endorse the full-blown version of the mental rules theory anymore. But this does not necessarily mean that the mental models approach has won the battle. As pointed out by Stenning (2002: 116 ff.), and others before him, the mental rules v. mental models debate is not likely to be resolved on empirical grounds,

---

[11] This is a reprint of some chapters from his 1994 book.

basically because the two approaches, in the simple cases studied (mainly syllogistic), are in fact formally equivalent. In such cases, there is a deductive (rule-based) system that is sound and complete with respect to a semantics, and thus 'reasoning processes modelled within one system can be emulated within each of the other systems' (Stenning 2002: 117). Ultimately, both approaches are a form of 'logicism',[12] i.e., the view that human reasoning is after all to be reduced to and measured against the canons of logic as traditionally construed.[13]

Some of the first researchers to contest the undisputed primacy of (classical) logic as providing the canons of correctness in reasoning tasks such as the selection task were Oaksford and Chater (1994). They argued that the experimental data should be reassessed in the light of a probabilistic, Bayesian model of optimal data selection in inductive hypothesis testing rather than according to 'a now outmoded falsificationist philosophy of science' (Oaksford and Chater 1994: 608). Against this different normative model, participants' performances are not nearly as catastrophic as they are according to the traditional, 'logical' model. Their analysis suggested that one could not simply take for granted either how participants actually interpreted the task at hand or that classical logic should necessarily provide the only standard of correctness in reasoning tasks.

More recently, several researchers have been investigating how people reason with conditionals within a probabilistic framework, – for example, David Over, Jonathan Evans, the Salzburg group led by Gernot Kleiter, among others (see Oaksford, Hahn, and Chater 2008 for a recent overview). One of the important discoveries has been that people tend to view the conditional probability as the probability of the conditional: the probability of 'if A then B' is typically interpreted as the probability of B given (the probability of) A, following the basic idea of the so-called Ramsey test for conditionals (Over and Evans 2003; Evans and Over 2004). The

[12] 'Logicism' tends to be understood by psychologists as pertaining to a Piagetian, 'mental rules' model of human reasoning. However, in current logic the rule-based approach, usually referred to as the proof-theoretic approach, is but one of the currently practised lines of inquiry. In fact, currently a different line, known as the model-theoretic approach, is arguably even more influential. So here I do not use the term 'logicism' only to refer to a 'mental rules' conception of reasoning; I use the term to refer to any approach to reasoning which claims that reasoning should conform (even if only in Marr's computational level) to classical logic.

[13] I will have more to say on mental models below. For now, let me quote a remark by Stenning and van Lambalgen (2008: 292) following their analysis of mental models theory: 'It will be clear from the above exposition that Johnson-Laird and Byrne have inadvertently tried to reinvent logic, while claiming to provide an alternative.' In effect, the very debate between mental rules and mental models approaches is now essentially outdated, as it has become clear that neither approach is particularly successful as a model of human reasoning.

probabilistic approach in investigations on human reasoning is now firmly established, and an important upshot is that it allows for easier communication with other fields where probabilistic (often Bayesian) frameworks are used, such as the recent field of formal epistemology. For our purposes, the importance of this 'probabilistic turn' in the psychology of reasoning tradition is that it illustrates once again the inadequacy of a traditional conception of logic as a model of human rationality, where the only two possible attitudes vis-à-vis a piece of information are outright acceptance or outright rejection (if a sentence is true or if it is false). In particular, these probabilistic frameworks are able to capture the strong component of defeasibility/non-monotonicity in human reasoning (to be discussed extensively in this chapter), which, by its very nature, the traditional deductive paradigm cannot capture.

4.1.2 *Context-sensitivity and interpretation of the task material*

Let us now go back to the original Wason selection task. As we have seen, content was found to have a facilitating effect in variations of the task, at least in the case of reasoning about towns and transportation. However, in similar experiments which used different themes as a source of content, the facilitating effect has not been observed (Manktelow and Evans 1979); this suggested that not just any arbitrary contentual version of the Wason selection task would produce the facilitating effect. Thus, researchers set out to identify what exactly had a facilitating effect, rather than just content.

One natural hypothesis was to consider that participants must also have some degree of familiarity with the content in question, so that prior experience could be activated. In the case of 'Every time I go to Manchester, I travel by train', presumably participants were well acquainted with the situation of travelling to different cities and taking different means of transportation. However, the conditionals used by Manktelow and Evans (1979), for example, did not contain particularly unknown content ('If I eat pork, then I drink wine'), and yet the facilitating effect was not observed. Why?

An experiment conducted by Johnson-Laird, et al. (Johnson-Laird, Legrenzi, and Legrenzi 1972) in the United Kingdom in the early 1970s, which could then not be replicated by Griggs and Cox (1982) in the USA in the early 1980s, provided researchers with further clues as to what factors seemed to produce a facilitating effect. The conditional used by Johnson-Laird, et al. was: 'If a letter has a second class stamp, it is left unsealed.' At the time, there was indeed postal legislation according to which a less expensive stamp could be used provided that the letter remained unsealed. Johnson-Laird and collaborators found that a very high percentage of participants gave the 'logically correct answer', namely to turn cards with letters having a second class stamp and cards with sealed envelopes. Later, Griggs and Cox (1982) observed no facilitating effect whatsoever in their US participants with this very conditional, and hypothesized that this was due to their lack of familiarity with the particular content. By contrast, a conditional expressing a rule that their US participants were presumably very familiar with led again to many more 'logically correct' responses: 'If a person is drinking beer, then the person must be over nineteen years of age.' In this case, 75 per cent of participants selected the 'right' cards, namely cards with a person drinking beer and cards with an age below nineteen. This could, in the first instance, be interpreted as a sign that familiarity with the content does indeed play a prominent role.

However, researchers soon realized that there was something normative, something 'rule-like' in many of the formulations of the selection task where a facilitating effect was observed. The stamp case referred to a UK postal legislation in force at the time; the drinking age case to a law of the state of Florida, where the experiment was conducted. Such conditionals are now often described as *deontic* conditionals, i.e., conditionals whose consequent describes how things *ought* to be if they are as described by the antecedent. By contrast, the original formulation of the selection task with numbers and letters simply describes a (putative) regularity. So a natural manipulation to be undertaken is to formulate a task with a deontic conditional with unfamiliar content, so as to probe whether familiarity or 'deonticness' is the decisive facilitating factor. Ultimately, this is already what had been done by Wason and Green (1984), but it was not immediately apparent that the conditionals with unrealistic or unfamiliar content ('Anyone consuming Coca-Cola on these premises must be at least 100 years old', 'Any lengths of red wool must be at least six meters long') were deontic conditionals. At any rate, the manipulations by Wason and Green (1984) elicited a much higher percentage of 'logical' responses than in the classical selection task, in spite of the displaced or unfamiliar content.[14]

Some years later, Cosmides and collaborators set out to probe the facilitating effect of formulating conditionals specifically with 'social contract' material. They concocted the scenario of a tribe on a remote tropical

[14] Evans (2007: chap. 4) distinguishes between the abstract–thematic axis and the orthogonal indicative–deontic axis in classifying facilitation effects on the Wason selection task. He argues that facilitation effects are restricted to conditionals which are both deontic and thematic.

island, the Kaluame, and formulated the task with what could be described as 'social contract material': 'If a man eats cassava root, then he must have a tattoo on his face.' Participants are then asked to verify (by the usual procedure of turning cards) whether four men of the tribe do or do not conform to this particular law governing rationing privileges (Cosmides 1989). This material, although obviously unfamiliar to participants, was found to elicit a high percentage of 'logical responses' (turning cards with cassava root and cards of men with no tattoo on their faces). Cosmides and collaborators went on to argue for the extravagant thesis of an innate, evolutionarily selected-for 'cheating mechanism', thus postulating that we are particularly good at reasoning in the specific domain of social contracts.[15] Regardless of the conclusions they drew from the experimental results, the results themselves did reveal something important, namely that the facilitating effect of deontic conditionals can be observed even with unfamiliar content, as had already been suggested by the results of Wason and Green (1984).

Perhaps the main lesson to be drawn from these results concerns the notion of 'logical form'. I have already mentioned that psychologists typically rely on a fairly naive understanding of the concept: the logical form of a sentence or argument would straightforwardly be 'read off' from its surface structure. But, at least since Russell, most philosophers are well aware of the fact that the 'logical form' of a sentence, if there is indeed such a thing, is often not straightforwardly correlated with its surface, grammatical structure. More importantly, what the experiments with the postal conditional illustrate is that logical form is not something that a sentence or argument *has*, in an independent, quasi-metaphysical sense; rather, logical form is at best something that speakers *attribute* to sentences or arguments by means of an interpretation. Thus, and in line with arguments I offered in Dutilh Novaes 2012, I maintain that it is best simply to stop using the concept of 'logical form' as a property of sentences and arguments. Even understood as something that is attributed to sentences and arguments by speakers, the concept evokes too many infelicitous connotations. Rather, it seems more appropriate to speak more generally of the *semantic interpretation* given to a sentence by a speaker, so as to avoid conceptual muddle.

[15] It was also part of the argumentation to argue against the role of 'logical reasoning' in human cognition insofar as the latter is purportedly 'topic-neutral'. But the topic-neutral characterization of logic, although influential throughout its history, no longer applies to many of the important recent developments in logic. Again, it is a case of psychologists dismissing the role of logic in human cognition for the wrong reasons, i.e., in virtue of a limited conception of logic.

Indeed, on the basis of their background knowledge of the relevant postal regulations, British participants attributed a deontic reading to 'If a letter has a second class stamp, it is left unsealed'; North American participants, in contrast, lacked background knowledge and interpreted the same conditional as a descriptive conditional. So the facilitating effect is related not to the very formulation of the conditional but to how participants interpret it. Stenning and van Lambalgen (2008) have emphasized the role of *reasoning to an interpretation* in the process of tackling a reasoning task, correctly pointing out that so far researchers have mostly focused on the *reasoning from an interpretation* part of the task (apparently assuming that the to-an-interpretation part would be unproblematic and uniformly carried out). Given sufficient elements, a participant can be induced to give a deontic interpretation to a conditional even if she would otherwise have no background knowledge of the scenario (as in Cosmides's experiment).[16] Hence, rather than eliciting prior experience as such, one important facilitating effect seems to be if participants assign a deontic interpretation to the conditional.

But why is reasoning with deontic conditionals so much easier for participants? Stenning and van Lambalgen (2004; 2008) have written extensively on the topic, and their views seem to me to be essentially correct. They say:

Our proposal about the selection task at its simplest is that the semantic difference between descriptive and deontic conditionals leads to a processing difference when these conditionals are used in the selection task. In barest outline, the semantic difference is this. Deontic conditionals such as the drinking age law cannot be false, they can only be violated. Hence turning a card never affects the status of the rule. Now whether a given customer violates the drinking age law is independent of whether another customer does so. This seems a trivial observation, but a comparison with the case of descriptive rules will show its importance. A descriptive rule can be true or false, and here the purpose of turning cards is to determine which is the case; in this case, unlike the previous one, the result of turning a card can affect the status of the rule. (Stenning and van Lambalgen 2008: 51)

In other words, with a deontic interpretation, the task arguably acquires a structure that is easier to process by the participant, and this would explain why the results in such cases are so different. This observation is also connected to another feature of how participants typically interpret a conditional: the fact that they often seem to interpret a conditional not as

---

[16] There have also been experiments with formulations of the conditionals explicitly with normative terms, such as, 'If a letter is sealed then it *must* carry a 20 cent stamp on it' (Cheng and Holyoak 1985).

a material implication but rather 'as a defeasible rule robust to exceptions' (Stenning and van Lambalgen 2008: 48). Deontic conditionals have precisely this characteristic: they are not (necessarily) defeated in case somebody does not comply with the rule. Moreover, descriptive conditionals are also typically interpreted as statements of regularities, which could, in principle, be falsified by a counterexample: 'If it rains, I will get wet' seems like a perfectly fine conditional statement, even though there are many circumstances which would count as counterexamples (e.g., I have an umbrella). In the original formulations of the selection task, however, participants were expected to treat the conditional as a material implication, not admitting exceptions or counterexamples, and this is one of the reasons why the experimenter's expectations and the participant's interpretation of the task are clearly in mismatch.

The probabilistic account of conditionals mentioned above has in common with Stenning and van Lambalgen's approach the fact that it treats conditionals as defeasible. To be sure, there are important differences between the two frameworks, but the defeasibility of conditionals is crucial to both. The greater success of both frameworks in accounting for the empirical data (relative to the original conceptualizations based on classical logic) again corroborates the inadequacy of expecting participants to interpret conditionals as indefeasible material implications.

4.1.3 *Non-monotonicity*

The defeasible interpretation of the conditional leads us directly to one of the most fundamental aspects of mismatch between the canons of deductive reasoning and how human agents indeed seem to reason: the issue of monotonicity v. non-monotonicity. As is well known, one of the key characteristics of a deductively valid argument is its monotonicity and indefeasibility. If d follows deductively from a set of sentences K, then it follows deductively from any set K' resulting from the addition of extra sentences to K (monotonicity). Similarly, if d follows deductively from a set of sentences K, then no extra premise added to K will ever bring about that d does not follow deductively from K', if K is a subset of K' (indefeasibility). So deductive validity is construed as indifferent to the addition of extra information (Oaksford and Chater 2002: sect. 2); if it holds, it holds *come what may*. Conversely, to show that a given argument is deductively *in*valid, it is sufficient to provide one single counterexample, one single situation (no matter how far-fetched it is) where the premises are the case and the conclusion is not the case.

Now, as anyone who has taught an introductory logic course knows, it is typically very hard for students to absorb the idea that one single counter-example invalidates an argument. They seem to come equipped with the idea that, if the conclusion is true when the premises are true in *most* cases, or at any rate the most *plausible* ones, then it is a good argument.[17] In other words, a fundamental characteristic of how people reason and argue outside of specific contexts such as mathematics seems to be captured by the idea of defeasibility and non-monotonicity. Naturally, this is a claim that requires further (empirical) corroboration, and here I will only present some of the results and arguments that seem to point in this direction. Nevertheless, we have already seen that one of the possible reasons why participants 'fail' to provide the 'logical response' in the case of descriptive selection tasks is that it is rather unnatural for them to interpret the conditional as the indefeasible material implication. So the matter of (in)defeasibility is present both on the object-language level of how the conditional is interpreted and on the meta-level of how the validity of an argument is evaluated (which is not surprising, given the tight structural connection between conditional sentences and arguments).

One experiment whose results suggest that patterns of human reasoning have a non-monotonic component is Byrne's (1989) 'suppression task'. Byrne herself did not have non-monotonicity or defeasibility in mind at all when designing the experiment or interpreting the results, but her results are readily interpreted from the point of view of these concepts.

Participants were presented with the following premises: 'If she has an essay to write she will study late in the library. She has an essay to write.' In this case roughly 90 per cent of participants drew the conclusion 'She will study late in the library.' This seems like good news to the partisan of a logicist view of human reasoning: the great majority of participants are able to perform a simple instance of *modus ponens*. However, when an additional premise was added, namely 'If the library is open, she will study late in the library,' in the very same group of participants now only 60 per cent drew the conclusion 'She will study late in the library'; so 30 per cent of them retracted their original conclusion. In first instance, this seems like a breach of monotonicity: if the argument from the original premises to the conclusion was valid, then it should have remained valid even with the addition of an extra premise. Equally puzzling is the fact that when a different extra premise was added to the original pair of premises, namely 'If she has a textbook to read, she will study late in the library,' then 94 per cent of the

---

[17] This intuition is aptly captured in probabilistic accounts of reasoning.

participants drew the original conclusion.[18] Why is it that the first additional premise defeats the argument for 30 per cent of the participants, while the second additional premise, seemingly 'identical' in 'logical form' (they are both conditional sentences) does not?

Byrne took her results to support the mental models account of human reasoning, because the latter can explain 'how premises of the same apparent logical form can be interpreted in quite different ways'. According to her,

The process of interpretation has been relatively neglected in the inferential machinery proposed by current theories based on formal rules. It plays a more central part, however, in theories based on mental models. (Byrne 1989: 83)

Of course, once one discards the assumption that 'logical form' is a feature inherent to sentences and arguments, then the issue can be formulated in very different terms. What Byrne seems to understand by 'logical form' here is simply the superficial grammatical structure of the sentences in question.[19] But, as pointed out by Stenning and van Lambalgen (2008: sect. 7.1), it would be more accurate to say that the sentence itself receives different interpretations (different attributions of 'logical form') rather than to say that the logical form of a sentence may be differently interpreted. My own view of Byrne's interpretation is that she is right about the negative effects of the results (to counter the mental rules account of reasoning), but the results do not seem to lend additional support to the mental models account.

At any rate, Byrne's results do suggest something important about human reasoning, namely the fact that we typically view arguments as defeasible. The effect is so subtle that even the addition of a conditional rather than a categorical premise may have the effect of compelling us to retract a given conclusion. In the case mentioned above, the premise which caused the 'suppression effect' is the conditional 'If the library is open, she will study late in the library,' and thus not the categorical 'The library is closed,' as could have been expected. In fact, if the added premise was 'The library is open,' most likely the suppression effect would not have taken place, for obvious reasons. What does the additional conditional premise add to the participant's reasoning process? Well, what it seems to add is the possibility that the library *might* be closed, something that had not been

---

[18] Stenning and van Lambalgen's data (2008: 160).
[19] As already mentioned, most philosophers would stress the distinction between surface structure and logical form, but many of them still seem to view the logical form of a sentence as a property it has independent of specific interpretations. Now, as the postal version of the selection task suggests, this assumption is simply unwarranted.

under consideration thus far; and if the library is indeed closed, then, no matter how many essays to write she has, she will *not* study late in the (closed) library. In the second case, however, the addition of the premise 'If she has a textbook to read, she will study late in the library' does not bring in any new elements which might prevent the conclusion from coming about. Having or not having a textbook to read will not affect her studying late in the library if she *also* has an essay to write. Thus, clearly not just any added premise may lead a participant to revise her previously drawn conclusion; this will only happen when the added premise brings in a new element which may affect the realization of the situation described in the conclusion. This only goes to show that, when drawing *modus ponens* in the simple case (with just two premises), participants seem to take the conditional in question to be 'a defeasible rule robust to exceptions', which in turn means that the whole argument, while 'valid' in some sense of 'valid' (perhaps something like 'sufficiently reliable'), is not deductively valid because it can be defeated by additional information.

Stenning and van Lambalgen (2008) model the reasoning processes that seem to be taking place in such cases in terms of 'closed world reasoning', a framework originally developed within artificial intelligence, and which they put to use to describe actual reasoning patterns of human reasoners. Crucially, closed world reasoning is non-monotonic, so Stenning and van Lambalgen's apparent success in capturing at least some fundamental properties of human reasoning (and in accounting for many of the already available experimental results, such as Byrne's) lends support to the claim that human reasoning is profoundly non-monotonic. And if this is indeed the case (as I think it is), then a traditional conception of deduction and logic simply cannot offer an accurate, universal/monolithic account of how humans reason, given the absolutely crucial role that the concept of monotonicity occupies in the traditional conception of deduction and the ubiquity of non-monotonic patterns in human reasoning.

I will have more to say on non-monotonic logics and closed-world reasoning below, but let me just present the main lines of Stenning and van Lambalgen's formal interpretation of the conditional as a defeasible rule robust to exceptions. Take the conditional 'If you turn on the switch, then the light will go on'; prima facie, this sounds like a perfectly legitimate, 'true' conditional, the sort of thing one may say in real-life situations (although perhaps a bit on the trivial side). But, of course, it does not mean that in every single case where the antecedent is satisfied the light will go on; funny things can happen – for example, the light bulb is broken, the electric power has been cut off, etc. But these somewhat unlikely scenarios do not change the fact that the original conditional seems perfectly acceptable (i.e., we attach a high degree of belief to it); these counterexamples do not defeat it. One way of articulating what seems to occur is to identify a 'hidden premise' in the argument, namely something like 'nothing funny is going on'. So, if we were to spell this out in detail the conditional becomes 'if you turn on the switch and nothing funny is going on, then the light will go on.' Hence, one way to formalize this conception of a conditional is as a ternary connective: $p \wedge \neg ab \rightarrow q$ (Stenning and van Lambalgen 2008: 178), where $\neg ab$ stands for 'nothing abnormal going on'. In other words, the semantics of conditionals thus understood assumes background information on how things are besides what is explicitly expressed in the antecedent and the consequent.

Notice, however, that, by making this so-called 'hidden premise' explicit, the conditional thus formulated in fact becomes an 'ordinary' material implication.[20] By making the parameter of (ab)normality explicit, the formalism restores the traditional 'logical' properties of the conditional. In effect, this is what happens more generally with so-called 'non-monotonic logics'. Strictly speaking, the term is a misnomer, as the logics themselves, i.e., the formal apparatuses, are entirely monotonic; they are non-monotonic in that they *model* non-monotonic reasoning, but one of the tricks for the modelling to be successful is precisely to internalize and make explicit the external parameters involved in the reasoning processes in question.

A hallmark of deductive reasoning is that no assumptions besides those explicitly formulated must be taken into account; no external information should be considered. Now, ordinary reasoning appears to be highly non-monotonic precisely because external information and background assumptions do play a significant role in reasoning processes; what non-monotonic logics do is to integrate these 'hidden assumptions' into the formalism itself, so as to make them explicit and tractable. Notice that this is not in any way intended as a criticism of these systems as models of the psychological, empirical property of the defeasibility of human reasoning: that the formalism has the technical, logical property of monotonicity does not mean that it cannot offer a suitable model of the psychological property of defeasibility (which is closely related to that of non-monotonicity). These are two distinct levels, and the fact that so-called non-monotonic logics are in fact monotonic formalisms simply shows that, on the theoretical, meta-level, monotonicity as a technical property has many benefits.[21]

---

[20] Compare with the concept of 'hidden premises' in the case of enthymemes.

[21] So, in a sense, one could say that non-monotonic logics are good models of non-monotonic reasoning on a general functional level (Marr's computational level), but not on an algorithmic level.

However, while I do have a great deal of sympathy for Stenning and van Lambalgen's characterization of the conditional as a ternary connective, I think it is important to keep in mind that, in actual reasoning processes (i.e., not in the formal models thereof), the 'nothing funny is going on' premise most likely maintains its status of tacit, background assumption. Arguably, it is not a piece of information that the agent herself is actively aware of.[22] Indeed, when it is said that the conditional is typically interpreted as 'a defeasible rule robust to exceptions', what the agent has in mind is arguably the binary connective $p \rightarrow q$, and the clause to the effect that nothing abnormal is the case is only tacitly assumed. Notice in particular that the ternary connective $p \wedge \neg ab \rightarrow q$ does not have an abnormal situation where the antecedent is the case and the consequent is not as an 'exception' (counterexample), simply because in this case one of the conjuncts of the antecedent ($\neg ab$) is falsified, and thus the whole antecedent is falsified; thereby the conditional simply behaves like a regular material implication, which is not falsified by a $F \rightarrow V$ valuation.

Hence, it is precisely the role and shape of this background information, which seems to be tacitly assumed in reasoning processes, that must be investigated, as we will do in the next section.

4.2 'A FUNDAMENTAL COMPUTATIONAL BIAS'

At the end of Chapter 2, I briefly mentioned what Stanovich refers to as a *fundamental computational bias* in human cognition. Let me quote (more fully) the key passage again:

The rose problem [*to be discussed shortly*] illustrates one of the fundamental computational biases of human cognition: the tendency to automatically bring prior knowledge[23] to bear when solving problems. The fact that prior knowledge is implicated in performance on this problem, even when the person is explicitly told to ignore the real-world believability of the conclusion, illustrates that this tendency toward contextualizing problems with prior knowledge is so ubiquitous that it cannot easily be turned off – hence its characterization here as a fundamental

---

[22] Keith Stenning points out that the introduction of awareness into the discussion may end up being counterproductive given that, to this day, there are no satisfactory psychological models of awareness in reasoning, which would probaby make most of the data uninterpretable. This may be so, but I still believe that maintaining awareness in our conceptual, theoretical horizon is important, even if empirically it is unclear how predictions involving awareness could be tested.

[23] Technically, it is not a matter of *knowledge* as philosophers understand it, i.e., as involving factuality, but rather a matter of *belief*, given that a participant will act in the exact same way towards her false beliefs.

computational bias (one that pervades virtually all thinking, whether we like it or not). (Stanovich 2003: 292–3)

First, let me say a few words on the very concept of reasoning and cognitive biases, as used in the psychology of reasoning tradition. As we have seen, participants typically do not do well in experiments with deductive tasks; but the 'mistakes'[24] they make are not random, rather they reflect systematic patterns that have been identified in the literature. To account for the possible mechanisms underlying this systematicity, the notion of 'biases' has been introduced. The different concepts of biases can be viewed as attempts to understand why participants systematically deviate from 'normative responses' in experiments with deductive tasks.

The term 'bias' has a strong negative connotation, and when it was first introduced in the psychology of reasoning literature to account for the systematic discrepancies between participants' performances and the normative responses in the experiments, the intended meaning was indeed that of an error, a cognitive vice. Nowadays, however, the concept of biases is much more nuanced, and several authors acknowledge that there are reasons to view many of these patterns as being perfectly beneficial in most situations.

It is also useful to distinguish several different meanings of 'bias'. In one, evaluatively neutral sense, bias refers to a tendency or inclination . . . Bias can also refer to a systematically flawed judgment process that is ultimately deleterious to the interests of the actor or society (as in 'racial bias'). Finally, a sort of middle ground has been suggested . . . akin to Simon's (1957) concept of bounded rationality. People may deviate systematically from theoretical standards, but may still be behaving optimally when broader concerns are taken into account (mental effort, cognitive capacity, emotional well-being, multiplicity of goals, etc.). (Klayman 1995: 386)

So, in order to avoid misunderstandings, let me state at the outset that I endorse the view that many such biases are essentially reasonable cognitive tendencies, certainly in most everyday-life, practical situations. In fact, I will argue that the theoretical standards encapsulated by the deductive canons are rather contrived (section 4.3); their normative import should not be extended beyond specific situations, namely situations of scientific inquiry. For most everyday-life situations, to reason purely deductively would lead to disastrous consequences, mostly (but not exclusively) for

---

[24] There is quite a bit of discussion in the literature on whether the discrepancies between participants' responses and the normative responses can be characterized as 'mistakes' at all, given that, at least on some interpretations, perfectly reasonable cognitive patterns are in fact being deployed.

reasons related to the concept of bounded rationality. The real challenge for a human reasoner is to select the truly relevant bits of information for the task at hand, amidst a situation of information overflow.

Consider the very bias we will focus on in the next pages, namely the tendency to bring prior beliefs to bear when reasoning about specific situations; in most cases, it would be utterly absurd *not* to take into account the information one already possesses when reasoning about a certain issue (cf. section 4.3.1). The Bayesian framework has even incorporated what Stanovich describes as a 'bias' into its formalism, in the form of the 'prior' (shorthand for 'prior probability'). Nevertheless, there are reasons to believe that at least in some situations, in particular scientific situations, the tendency systematically and inadvertently to bring prior beliefs to bear when reasoning is not entirely advantageous. I now review some of the empirical findings suggesting that this is indeed a pervasive reasoning tendency in human cognition.[25]

4.2.1 *The belief bias effect*

A seminal study of participants' tendency to reason towards confirmation of the beliefs they already hold (the first to present robust results) is Evans, Barston, and Pollard (1983).[26] In this paper, and in many subsequent papers, the issue is presented in terms of a 'conflict' between logic and belief, and the basic reasoning mechanism to be investigated is the fact that participants 'will tend to endorse arguments whose conclusions they believe and reject arguments whose conclusions they disbelieve, irrespective of their actual validity' (Evans, Barston, and Pollard 1983: 295). This tendency has received the name of 'belief bias' in the literature.[27] The experiments in this study as well as in the vast majority of subsequent studies are formulated with simple syllogistic arguments.[28] More generally, even if there are methodological and conceptual problems with the experimental framework, the results are so robust that they cannot be dismissed simply on account of these problems.

---

[25] Evans (2007: chap. 4) provides an overview of the effect of content and belief in reasoning.

[26] There had been some older studies (in the 1940s) identifying the effect of prior belief in reasoning processes, but Evans, Barston, and Pollard (1983) revived the debate.

[27] There is another closely related concept, the concept of 'confirmation bias' (Nickerson 1998). Confirmation bias is usually intended as a more general phenomenon, whereas belief bias concerns specifically reasoning processes.

[28] While one may think that the limited range of the task materials may suggest that the tendency is not as general as it is presented to be, I firmly believe this not to be the case. The phenomenon identified seems to be perfectly general.

In the study, participants were given fully formulated syllogistic arguments (two premises and one conclusion) and asked to evaluate their
validity. Among the syllogisms presented to participants, some are valid,
some are invalid, some have a believable conclusion, and some have an
unbelievable conclusion (in the four possible combinations). Table 4.1
contains some of the syllogisms from Evans, Barston, and Pollard 1983;
notice that the valid/invalid syllogisms (both believable and unbelievable)
are of the same syllogistic figure and mood respectively. Table 4.2 indicates the general results of the experiment (percentage of arguments
accepted as valid).

Clearly, prior beliefs seem to be activated when participants are evaluating (the correctness of) arguments. Participants can be seen as showing a
degree of what has been called 'logical competence' (Macnamara 1986),
assuming the traditional canons, in that valid syllogisms are more often
endorsed than invalid syllogisms in both categories. Moreover, syllogisms
with believable conclusions are more often endorsed than syllogisms with
unbelievable conclusions. What is remarkable though is that *in*valid syllogisms with believable conclusions are more often endorsed (71 per cent)
than valid syllogisms with *un*believable conclusions (56 per cent).

Table 4.1

Valid-believable	Valid-unbelievable	Invalid-believable	Invalid-unbelievable
No police dogs are vicious.	No nutritional things are inexpensive.	No addictive things are inexpensive.	No millionaires are hard workers.
Some highly trained dogs are vicious.	Some vitamin tablets are inexpensive.	Some cigarettes are inexpensive.	Some rich people are hard workers.
Therefore, some highly trained dogs are not police dogs.	Therefore, some vitamin tablets are not nutritional.	Therefore, some addictive things are not cigarettes.	Therefore, some millionaires are not rich people.

Table 4.2

	Believable conclusion	Unbelievable conclusion
Valid	89	56
Invalid	71	10

Notice also that the effect of the believability of the conclusion is most acutely felt in the case of *in*valid arguments: a 61 per cent difference over a 33 per cent difference between endorsed valid arguments having a believable conclusion and endorsed valid arguments having an unbelievable conclusion. Here, we are interested both in the patterns leading participants to endorse invalid arguments with believable conclusions and in the patterns leading participants to reject valid arguments with unbelievable conclusions, but it is important to notice that the effect is most pronounced in the first case (71 per cent of the responses were 'logically' incorrect, as opposed to 44 per cent in the case of valid arguments with unbelievable conclusions).[29]

Naturally, one concern that can be raised regarding such experiments is whether participants do in fact understand what is asked of them, and in particular whether they understand the concept of the *validity* of an argument. Validity is obviously a highly technical, theory-laden concept, and one of the methodological faults of such experiments seems to be to assume that participants will be thoroughly familiar with the concept and give it a univocal interpretation. The exact formulation of the task as presented to participants in the Evans, Barston, and Pollard 1983 study was as follows:

This is an experiment to test people's reasoning ability. You will be given four problems. In each case, you will be given a prose passage to read and asked if a certain conclusion may be *logically deduced* from it. You should answer this question on the assumption that all the information given in the passage is, in fact, true. If you judge that the conclusion *necessarily follows* from the statements in the passage, you should answer 'yes,' otherwise 'no.' (Evans, Barston, and Pollard 1983: 298; emphasis added)

It is of course far from obvious that participants will understand the concepts of 'logically deduced' and 'necessarily follows'. Participants in such experiments are typically undergraduate students, thus people with a fair amount of formal education, but they are explicitly screened so as to exclude those with specific training in logic. But what do these concepts mean to people without training in logic? Some of them may have some indirect acquaintance with such concepts, but it is fair to assume (although this is obviously an empirical question) that many of them will have very imprecise and probably off-the-mark interpretations of this terminology.[30]

[29] I owe this point to Shira Elqayam.
[30] Morris and Sloutsky (1998) investigate how mastery of the concept of logical necessity develops upon extensive training in mathematics.

Moreover, even if we suppose that participants typically interpret the instructions as asking them to judge whether an argument is 'good', 'compelling', 'reliable', 'correct' (in non-technical senses), it is still noteworthy that they tend to endorse arguments whose conclusions accord with their prior beliefs.

It may also be objected that, while this form of attachment to prior beliefs – which can also be described as doxastic conservativeness – may be operational when participants are asked to evaluate the correctness of fully formulated arguments, it may not have the same effect when people are actively engaged in reasoning, i.e., in drawing conclusions from premises themselves. But as it turns out, the belief bias effect has also been observed in conclusion production tasks, such as in Oakhill and Johnson-Laird 1985.

Indeed, Oakhill and Johnson-Laird (1985) present compelling evidence supporting the claim that belief bias affects not only the evaluation of the correctness of arguments, but also the very formulation of arguments and the process of drawing conclusions. In their study, participants were presented with pairs of premises and asked to draw conclusions: the premises were carefully selected so as to control for different variables. They used two pairs of schemata, one of which leads to exactly one conclusion according to the syllogistic standards of validity (following the valid AOO-2 mood), while the other one does not:

Some of the A are not B.	Some of the A are not B.
All of the C are B.	All of the B are C.
(∴ SOME OF THE A ARE NOT C.)	(NO VALID CONCLUSION.)

Using these two pairs of schemata, they formulated different pairs of premises. For the first one (the one from which a conclusion can be deductively derived), pairs of premises were constructed with different contents: contents that led to a believable conclusion (the first example below) and contents that led to an unbelievable conclusion (the second example below).

Some of the healthy people are not vegetarians.
All of the athletes are vegetarians.
(∴ SOME OF THE HEALTHY PEOPLE ARE NOT ATHLETES.)

Some of the athletes are not vegetarians.
All of the healthy people are vegetarians.
(∴ SOME OF THE ATHLETES ARE NOT HEALTHY PEOPLE.)

For the second one, i.e., the pair of premise-schemata that does not allow for a deductively valid inference, a similar procedure was adopted: pairs of premises were constructed with different contents. In this case, the 'conclusions'[31] aimed at were instantiations of the schema 'Some A are not C,' which is frequently (but erroneously) produced as a conclusion to pairs of premises of this kind. Again, some contents would induce the 'conclusion' of a believable sentence, while other contents would induce the 'conclusion' of an unbelievable sentence.

Some of the women are not beautiful.
All of the beautiful people are actresses.
(Expected 'conclusion': some of the women are not actresses – believable.)

Some of the actresses are not beautiful.
All of the beautiful people are women.
(Expected 'conclusion': some of the actresses are not women – unbelievable.)

Another factor taken into account in the study is whether the potential conclusion (either validly or invalidly inferred) was (in the authors' terms) definitionally or empirically false. 'Some of the actresses are not women,' is definitionally false; 'Some of the athletes are not healthy,' is (presumably) empirically false.[32] With these three parameters (believability of the conclusion, validity of the argument, conclusion being definitional or empirical), eight pairs of premises were produced. At the experiment, participants were given these eight pairs of premises (plus three filler items), shown the following possibilities, and asked to indicate what could be 'deduced with absolute certainty' from each pair of premises:

All of the ____ are ____.
Some of the ____ are ____.
None of the ____ is a ____.
Some of the ____ are not ____.
No valid conclusion.

The results are spelled out in detail in the article (Oakhill and Johnson-Laird 1985: 561), but let me summarize the main points. With pairs of premises that do lead deductively to a conclusion (forming a AOO-2 syllogism), the majority of participants (58 per cent) correctly draw an empirically false unbelievable

---

[31] I will use scare quotes to denote that the 'conclusion' in question cannot be validly inferred from the premises, and thus constitutes an inferential error from the point of view of syllogistic.

[32] This is of course not obvious, given the number of athletes that make use of unhealthy illegal substances to enhance their performances.

conclusion (only 13 per cent state that there is no conclusion to be drawn). But if the correct conclusion is definitionally false (and thus even less believable), then only 38 per cent of the participants draw the conclusion, while 46 per cent state that there is no conclusion to be drawn. In other words, it might have been expected that the belief bias effect would have led participants to reject the correct conclusion also if it is merely empirically false, but this has not been observed; however, in the case of highly unbelievable conclusions (definitionally false), the belief bias effect was clearly observed.

With pairs of premises that do not lead deductively to a conclusion, the belief bias effect was also observed. In the case of a believable putative 'conclusion' of the form 'Some A are not C,' a very low number of participants gave the correct answer, i.e., no conclusion to be drawn (17 per cent for one pair of premises and 21 per cent for the other): clearly, the appeal of the believable 'conclusion' was quite strong. In the case of the definitionally false putative 'conclusion', 50 per cent of the participants gave the correct answer, i.e., no conclusion to be drawn; but in the case of the empirically false putative 'conclusion', a lower number of correct responses (no conclusion to be drawn) was observed (29 per cent).

These figures, although not as univocal and clear as in the Evans, Barston, and Pollard (1983) study, do suggest that participants typically let their prior beliefs affect their conclusion-drawing. They often draw 'conclusions' that do not follow deductively but which are plausible/believable, just as they often refuse to draw a deductively inferable conclusion if it is implausible/unbelievable.

Given that the studies discussed so far deal with conclusions (or 'conclusions') which are either believable or unbelievable, a natural question to be asked is what would happen if the conclusion (or 'conclusion') is neither believable nor unbelievable. This is exactly what Sá, West, and Stanovich (1999) set out to do. They designed an experiment with so-called 'unfamiliar content', so that the tendency towards doxastic conservativeness would presumably not be activated. The prediction was that, if it was indeed doxastic conservativeness that was trumping inferential competence, in the case of syllogisms with unfamiliar content this would not occur, and participants would give 'logical' responses more often. In first instance, participants were given an invalid syllogism with familiar content, and in fact with a believable conclusion:

> All living things need water.
> Roses need water.
> Roses are living things.

As could have been anticipated, only 32 per cent of the participants gave the logically 'correct' response when evaluating this syllogism, i.e., that it is invalid. They were then given a little scenario of a different planet, involving an imaginary species, wampets, and an imaginary class, hudon, and subsequently were asked to evaluate the following syllogism:

> All animals of the hudon class are ferocious.
> Wampets are ferocious.
> Wampets are animals of the hudon class.

Interestingly, 78 per cent of the very same participants whose great majority had failed to give the 'logically correct' response in the previous task gave the 'logically correct' response here, i.e., that the syllogism is invalid. Even more significantly, the two syllogisms have the exact same figure and mood, AAA-2 (the universal quantifiers are omitted in the second premise and in the conclusion). So, while they had failed to 'see' the invalidity of the first syllogism, arguably in virtue of the familiar content, in the second case the unfamiliar content apparently made it so that prior beliefs did not interfere in the participants' reasoning. Stanovich (2003) refers to this phenomenon as 'the rose problem' (see quotation at the beginning of section 4.2), and experiments such as this led him to conclude that being tendentious toward prior belief is a 'fundamental computational bias' of human cognition.

For our purposes, it is crucial to reflect on what seems to be going on in the second case, i.e., the wampet/hudon argument. When discussing the Wason selection task, we have seen that participants are typically not very comfortable with reasoning with so-called 'abstract' material, i.e., with letters and numbers (which again should come as no surprise to anyone who has ever taught introductory logic classes and had to teach students to operate with 'contentless' formalisms). However, we have also seen that, when reasoning with 'contentful' material, participants are prone to belief bias. The wampet/hudon argument is remarkable in that it does not ask participants to operate with a technique that they are not familiar with, but at the same time it somehow manages to counter the effect of belief bias on reasoning competence. I submit that this occurs in virtue of the fact that the key terms in the argument (hudon, wampet) are *meaningless* to the participants, so effectively we have a process of 'de-semantification' – a notion already introduced in Chapters 2 and 3, but which will be extensively discussed in Chapter 6. I will return to the idea of de-semantification as producing a debiasing effect, in Chapter 7.[33]

---

[33] Stanovich (2009) elaborates on the idea of cognitive decoupling, which bears some similarity to the idea of de-semantification, but is ultimately quite different.

Notice that, while the studies I have just discussed are not particularly recent, the literature on belief bias has continued to grow in recent years; the results presented here have been replicated several times and remain essentially uncontested. Much of this literature focuses on details that are not directly relevant for the present discussion, but let me just mention a few of the recent studies which offer further insight into the patterns involved in belief bias. Torrens, Thomson, and Cramer 1999 explores individual differences and their effect on belief bias; Goel and Dolan 2003 investigates the functional neuroanatomy of belief bias; De Neys and van Gelder 2009 focuses on the development of the influence of belief bias across the lifespan of humans; Prowse, Turner, and Thompson 2009 argues that, while reasoners' confidence is poorly calibrated (they tend to be overconfident), training on the concept of logical necessity may improve calibration. In sum, the belief bias effect has been abundantly documented, and the details of the underlying mechanisms are still the object of much scientific interest.

4.2.2 *Bringing prior beliefs to bear: preferred models*

In their 1985 article, along with the results on spontaneous conclusion production, Oakhill and Johnson-Laird present the outlines of one possible account of how humans proceed when they reason (formulated as instructions for the reasoner):

1. Imagine a state of affairs in which the premises are true, i.e. construct a mental model of them.
2. Formulate a conclusion that is true in the model (or all the models that have been constructed), and that relates terms not explicitly related in the premises.
   If there is no such conclusion, respond that there is no valid conclusion.
3. If there is such a conclusion:
   Try to imagine a model of the premises in which the conclusion is false.
   If there is no such model, respond with the conclusion which is valid.
   If there is such a model, return to step 2. (Oakhill and Johnson-Laird 1985: 556)

These lines offer the gist of the mental models account of reasoning,[34] which is presented as an alternative to the mental rules 'Piagetian' account. The key idea is that reasoning proceeds not by the application of abstract rules, but rather by the representation of states of affairs. I think this

---

[34] Stenning and van Lambalgen (2008: 285) provide a similar description.

description is somewhat on the right track, but many important provisos should be made. For example, the reference to 'terms' in step 2 reveals that the approach, as formulated by Johnson-Laird and collaborators, is very language-dependent, formulated for the specific case of syllogistic reasoning. Moreover, the idea that the agent imagines one specific state of affairs in step 1 seems to me to be excessively precise; the agent may just as well imagine an arbitrary minimal state where the premises are true. Furthermore, the step-wise search for countermodels, which is an important feature of the mental models theory in that it is thought to yield empirical predictions, seems contentious; for reasons to be discussed in section 4.3.3, I believe that reasoners do not spontaneously search for countermodels. More likely, the agent simply represents one situation (or a class of situations) where the premises are the case (typically with the help of some background information concerning her environment) and checks what else is also the case in the situation. But, ultimately, while the exact details of the mental models approach as developed by Johnson-Laird and collaborators seem off the mark for a variety of reasons,[35] I do endorse the general 'hunch' that reasoning proceeds by representation of situations rather than by the application of 'mental rules'.

My reasons for endorsing this idea are of different kinds. First, from the point of view of philosophy of logic and recent discussions on the concept of logical consequence, I am convinced that the primitive notion of logical consequence is the one formulated in terms of necessity/impossibility: it is *impossible* for the antecedent to be the case while the consequent is not.[36] This can be interpreted in metaphysical as well as epistemic terms (i.e., pertaining to what an agent can represent), but, in both cases, proof-systems are a posteriori models of general patterns that emerge from the more primitive notion of logical consequence as necessary truth-preservation, i.e., monotonicity (this tendency can also be observed in the historical development of the notion of consequence).

Secondly, systems of inferential rules are highly language-dependent – or, in other words, presentation-dependent – and it is very hard to see how

---

[35] Stenning and van Lambalgen (2008: sect. 10.6) discuss in detail the many shortcomings of the mental models theory. Among other problems, its proponents simply do not have a worked-out account of the procedure(s) leading to the construction of a mental model, which is of course a fatal shortcoming in any theory purporting to explain how human agents reason in terms of the concept of mental models. Their criticism seems to me to discredit in a definitive way any claims that the mental models theory might have to being the correct account of the principles underlying human reasoning. Cf. Hodges 1993.

[36] See, for example, Etchemendy's (2008) defence of a 'representational' approach to logical consequence over an 'interpretational' one.

reasoning in general, in *all* humans, could possibly be guided by a specific system of 'mental' rules.[37] Here, I side with Stenning and van Lambalgen (2008: 144) in claiming that 'there is nothing particularly linguistic about logic', or more generally about reasoning and thinking. (This thesis has received ample empirical corroboration in recent decades.)

Thirdly, the true origin of the notion of logical consequence seems to me to be the body of *mathematical* practices (see section 4.3.3, below). Here again, explicit systems of inferential rules are only formulated a posteriori, and the great majority of practising mathematicians remain oblivious to the codifications of their inferential practices offered by different logical systems.

One may object that I am now betraying my own avowed conviction that (traditional) logic and the concept of deduction typically (though perhaps not always) do not correspond to the actual inferential practices of human agents outside specific contexts such as that of mathematics. Well, the point I am making here is rather that *even* in these highly theoretical, specific, and somewhat contrived contexts, systems of inferential rules occupy a secondary, derivative position. Therefore, it would be highly surprising if they would occupy a primary position in actual patterns of 'untrained' reasoning.

More importantly, the main support for the view that reasoning proceeds by representations of states of affairs comes from empirical evidence that *planning* is a fundamental cognitive feature of human beings, and one that we share with many species of non-human animals (which again reinforces the idea that, at its core, reasoning must not be language-dependent). I here understand planning in a very broad sense, also covering basic, possibly unreflective, actions such as motor actions, again following Stenning and van Lambalgen:

We plan with respect to *our expectations of the world*, not, as we would have to if we planned using classical logic, with respect to *all logical possibilities*. Maintaining a model of the current state of the immediate environment relevant to action is a primitive biological function; calculating what is true in all logically possible models of the current sense data is not. These planning logics are just as much what one needs for planning low-level motor actions such as reaching and grasping,

[37] Of course, this does not mean that one cannot adopt a particular proof-system to investigate psychological processes, to make experimental predictions, etc. But presentation-dependence is a feature inherent to any syntactic system, and one must always bear in mind that such presentation features are by no means necessarily present in the actual cognitive processes. Here again we have the distinct levels of the psychological, empirical phenomena to be studied and the technical aspects of the formalism used on the theoretical level.

as they are for planning chess moves. (Stenning and van Lambalgen 2008: 144; emphasis added)

So I take it that the best hypothesis we have so far concerning how human reasoning actually takes place is that it proceeds by mental representations of the relevant explicitly available information (in particular the premises) and by an inspection of what else seems to be the case (or would be the case) in such situations (pending, of course, a compelling account of the exact details of these representational skills). Notice that this general mechanism is also perfectly applicable in the case of counterfactual reasoning, i.e., taking as a starting point a situation that differs to various degrees from the actual situation the agent finds herself in, but the general principle would be the same.

In recent years, Evans (2006; 2007) has been developing a framework known as 'Hypothetical Thinking Theory' (HTT), and the account I propose here is in agreement with HTT in many respects. It is based on three general principles, all of which have close counterparts in the present account.

The theory assumes that reasoning and judgment are facilitated by the formation of epistemic mental models that are generated one at a time (singularity principle) by preconscious heuristic processes that contextualize problems in such a way as to maximize relevance to current goals (relevance principle). Analytic processes evaluate these models but tend to accept them unless there is good reason to reject them (satisficing principle). (Evans 2006: 378)

Evans's satisficing principle echoes the idea (*contra* the mental models approach) that reasoners do not readily look for counterexamples, and is closely related to Stanovich's fundamental computational bias. His singularity principle is related to the claim defended here that reasoners typically do not represent a large class of possibilities (although to say that they represent only one situation is a much stronger claim which I am not prepared to commit myself to). And his relevance principle might be seen as related to the notion of 'preferred models', to be presented shortly. But while Evans's principle explicitly makes reference to goals, being thus at least to some extent a pragmatic principle, the notion of 'preferred models' is essentially epistemic, and is related to our resistance to revising the beliefs we already hold (which is also captured in his satisficing principle).

The 'representational' approach to reasoning adopted here has the advantage of allowing for an explanation of the fundamental differences between actual patterns of human reasoning and the canons of deductive reasoning as traditionally construed. The key concepts are again those of monotonicity and non-monotonicity: while arguably the very goal of the canons of deductive reasoning as traditionally construed is to ensure monotonicity and indefeasibility, actual patterns of human reasoning are highly non-monotonic and defeasible (and for good reasons!). Let me spell this out in more detail.

As already mentioned, Stenning and van Lambalgen take the framework of closed world reasoning as developed within AI to be a good model of certain crucial aspects of human cognition (in particular what they call 'reasoning to an interpretation'); the cornerstone of this framework is known as the 'closed world assumption', as opposed to the 'open world assumption'.

Closed world assumption. Complete knowledge is assumed for the (typically small) portion of the world relevant in each reasoning situation. Concerning this portion of the world, unless a piece of information is known to be the case, it can be assumed not to be the case.

Open world assumption. We assume that our knowledge of the world, even of the small salient portion in question, might be incomplete, and therefore any information not explicitly specified is considered unknown.

In fact, we have already come across an instance of the closed world assumption when discussing conditionals interpreted as a ternary connective: the assumption that 'nothing funny is going on' is the closed world assumption for reasoning with abnormalities (Stenning and van Lambalgen 2008: 40). We assume that, in the absence of positive information to the contrary, nothing abnormal is going on.

Closed world reasoning can be described in syntactical terms, i.e., with a propositional language where the connectives behave differently from the classical connectives (e.g., the implication is construed as a ternary connective). In this case, the closed world assumption is rendered by the supposition that, unless an atomic sentence (or the negation of an atomic sentence) is known to be true, it is (assumed to be) false. Stenning and van Lambalgen offer extensive modelling of human reasoning in terms of this framework, but I take it that they do not mean to claim that the very syntactical rules described by the framework are actually and precisely implemented when people reason. Instead, as I read them, the formalism is presented as a *model* of the phenomena in question, just as a physical theory is a model of physical reality: an approximate description, not the 'real thing'.[38]

---

[38] In personal communication, Stenning clarifies that they take at least some aspects of the formalism to be accurate representations of psychological phenomena. For example, the formalism does presuppose an asymmetry between positive and negative information, and there are reasons to think that this asymmetry is a real psychological phenomenon (e.g., the discrepancy in reasoning competence with modus ponens v. modus tollens, which is naturally accounted for in terms of such an asymmetry).

However, given my own 'representational' view of reasoning, I believe that a semantic approach, even if still a model (in the sense of an approximate description), is a more accurate model of how humans reason. In particular, we can take the semantic definition of consequence as our starting point. Now, the standard definition of logical consequence dictates that one must inspect *all* logical possibilities, i.e., *all* models that verify the premises, in order to see what else holds in all these models and thus to be able to draw a conclusion:

(Logical) consequence. A sentence[39] φ is a logical consequence of a set of sentences Γ iff in all situations where all members of Γ are the case, φ is also the case.

But, clearly, looking into *all* logical possibilities is not in any way what humans seem to be doing when reasoning in most situations. Instead, they presumably take a small subset of situations where the explicitly available information (the premises) is the case, namely those situations that accord with additional background knowledge about the world – the agent's 'expectations of the world'. This general idea has been offered a precise technical formulation by Shoham (1987) in terms of *preferred models*. Shoham's goal was to offer a unified framework for the various non-monotonic logics proposed in the 1980s in the AI community. The notion of preferred models allows for the formulation of a non-monotonic consequence relation, but which is otherwise closely related to the traditional notion of logical consequence:

Preferential consequence: φ is a preferential consequence of Γ iff in all preferred models of Γ, φ is also the case.

Preferential consequence is non-monotonic because the preferred models of Γ', a proper superset of Γ, are not necessarily a subset of the set of preferred models of Γ (they may even be entirely disjoint). So, while φ may hold in all preferred models of Γ, the addition of an extra piece of information, say ψ, may make it so that φ does not hold in all preferred models of $\Gamma + \psi$, leading the agent to retract the previously drawn conclusion to the effect that φ.

The research programme on preferential consequence and preferred models has lost some of its original momentum in the AI community because, as it turns out, systems containing these notions often have

---

[39] I am using 'sentence' here because this is how the notion of logical consequence is often formulated. But, given the view that reasoning is not at heart a linguistic matter, it would be more accurate to offer a different formulation of the definition – for example, in terms of 'piece of information'. The technical term 'proposition', understood as abstract and non-linguistic, could also be used, and is indeed often used by those who take the relata in the relation of logical consequence to be non-linguistic objects.

extremely high computational complexity (and AI researchers are of course predominantly interested in systems with good computational properties that can thus be implemented). Nevertheless, as a conceptualization of what seems to be going on when human agents reason, the idea of preferred models seems to me to be essentially correct – although the formal modelling is still imperfect, as it assigns a very high computational complexity to a task that we seem to be able to accomplish with ease. What we do when we reason is often to take as our starting point representations of the current state of the immediate environment, in particular the elements that are relevant for the task at hand. Our 'preferred models' are thus precisely representations of our current state of belief, our current doxastic commitments, possibly updated by whatever counterfactual information we wish to take into account. It is no wonder that our reasoning processes are affected by the beliefs we hold, as observed in the belief bias experiments; in effect, they are to a great extent *guided* by the beliefs we hold. So the so-called 'conflict between belief and logic' should not be understood as a conflict between believing and reasoning, as reasoning *is*, at least in part, believing.

The notion of preferential consequence sheds new light on some of the results in the belief bias experiments. In the rose example, when many (78 per cent) participants judged that the inference from 'All living things need water' and 'roses need water' to 'roses are living things' is a legitimate inference, what might be happening is that, in all of the participants' preferred models where the premises are true (in fact, their preferred models *tout court*, as both premises are presumably held to be true by participants), the 'conclusion' is also true, simply because it is true in all of the participants' preferred models (they presumably all strongly believe that roses are living things). So this inference satisfies the definition of preferential consequence; the premises themselves do not rule out the possibility that roses do need water but are not living things, but such a situation is not among the reasoners' preferred models. It can be conjectured that, if another premise was added, namely 'Not all things that need water are living things', (some) participants might react by withdrawing their endorsement of the 'conclusion' (just as in the 'suppression task' experiment), as now they are made aware of this possibility.

In contrast, when judging the correctness of the inference from 'All animals of the hudon class are ferocious' and 'Wampets are ferocious' to 'Wampets are animals of the hudon class', participants could not resort to their own preferred models concerning the hudon class and wampets (as there was none), and therefore had to adopt a different reasoning strategy. Now, given that they were undergraduates with several years of formal education, they could probably resort to some of the alternative reasoning techniques they might have been exposed to in their school career (even if only implicitly), leading many of them to the judgment that this is not a correct inference. This is crucial: I conjecture that, if the same experiment was applied on a different population, and especially participants with no or little formal education, the results would be very different; see the results reported by Counihan (2008a).

However, explaining why participants reject a conclusion which nevertheless does follow logically from a set of premises on the basis of the notion of preferential consequence is not as straightforward. Notice that in Oakhill and Johnson-Laird's (1985) spontaneous conclusion production experiment, 46 per cent of the participants judged that no conclusion can be drawn from 'Some of the actresses are not beautiful' and 'All of the women are beautiful'. Thus, in particular, they rejected the conclusion 'Some of the actresses are not women', which, although definitionally false, does follow necessarily from the premises. More generally, any pair of premises and conclusion that satisfies the traditional definition of (logical) consequence in terms of *all* models *also* satisfies the definition of *preferential* consequence, as the set of preferred models of Γ is a subset (usually a proper subset) of the set of models of Γ. So, while the notion of preferential consequence elucidates why participants may draw inferences that are not deductively (logically) valid, it does not help elucidate why they often do not draw inferences that are deductively valid, in particular when the manifest conclusion does not accord with their prior beliefs.[40]

My tentative explanation for this phenomenon pertains to how we should understand humans' abilities to create representations of states of affairs. I have said that this general idea can accommodate counterfactual reasoning, i.e., humans can in principle represent states of affairs that differ to some extent from the current state of their immediate environment. It seems plausible, however, that there may be limits to these representational abilities: they can represent situations where not all actresses are beautiful and, by a stretch of the imagination, perhaps also situations where all women are beautiful. But 'merging' the two classes of

---

[40] The issue here seems to be the reasoner's success in 'blocking' real-world knowledge. There is a typical developmental course for children to become more proficient in this particular reasoning move (as suggested, for example, by the 'improved' performance on the false-belief task in children around four years old), but it seems plausible that there should be individual differences in this respect in adults. Indeed, while 46 per cent of the participants judged that there was no conclusion to be drawn in this case, 38 per cent of the participants *did* draw the counterintuitive conclusion.

situations is not entirely straightforward.[41] In practice, if they reject the belief that some actresses are not women, merging the two classes of situations would yield an empty class of preferred models, such that this particular reasoning mechanism breaks down.[42] Alternatively, the agent may represent inconsistent preferred models, where some actresses are not beautiful, all women are beautiful, and all actresses are women.[43] In other words, when there is indeed a conclusion that follows (deductively) from a set of premises but which is highly unbelievable to the reasoner, the pull of doxastic conservativeness may be so strong that she simply does not 'see' the conclusion.

In explicit deductive settings – for example, when doing logic or mathematics – we are taught that, when one (correctly) reaches a highly implausible conclusion from apparently sound premises, one of the premises must be rejected. Now, given the typical set-up of the experiments, i.e., given that participants are explicitly told to consider the premises as true, this option is not available to them, so a natural alternative to maintain one's core beliefs intact is to judge (erroneously) that there is no conclusion to be drawn from the premises.

In the next chapters, we will be mostly interested in cases of surprising, counterintuitive conclusions from given premises; we will be interested in patterns that may compensate for the tendency we have to simply 'fail' to identify a given counterintuitive conclusion that does follow (in one sense or another) from given premises. But, cases of overly optimistic conclusion-drawing, such as the rose example, must also be kept in mind, as patterns that might compensate for our tendency to draw believable 'conclusions' may also be required.

At any rate, I have argued that our tendency towards doxastic conservativeness, as evidenced by the belief bias effect, can be shed new light on by means of a representational account of reasoning mechanisms, and in particular by means of the notions of preferred models and preferential consequence.

---

[41] On integrating premises, see Andrews 2010.

[42] Notice that, thus formulated, preferential consequence does not exclude a somewhat counter-intuitive feature of the classical definition of consequence, namely the fact that if there is no model that satisfies the premises, then any conclusion follows (the *ex falso quodlibet* principle). Similarly, if there is no preferred model satisfying all the premises (although there might be a non-preferred model that does), then in principle any arbitrary conclusion would satisfy the definition of preferential consequence.

[43] As is now well known from research on paraconsistent logic and related fields, inconsistent sets of beliefs are not necessarily trivial sets of beliefs, where everything holds.

4.3 A PLURALISTIC CONCEPTION OF HUMAN RATIONALITY

In this section, I discuss in more detail the implications of the empirical findings just presented for a general conception of human rationality. As is widely known, these findings have sparked a major debate, the so-called 'rationality debate' (Elio 2002; Over 2003).[44] Given that the traditional (in fact, essentially Kantian) conception of human reasoning and rationality was closely related to the concepts of logic and deduction, the experimental findings suggesting that humans are not very proficient in reasoning deductively led many researchers to conclude that human beings are 'irrational'. Others have concluded instead that it is logic and deduction, in particular in that they are (purportedly) domain- and context-independent systems of reasoning, that have no natural place in human cognition, leading to the concept of 'ecological rationality' (Cosmides 1989; Gigerenzer 2007). These are essentially the two extremes in a spectrum of different positions that have been defended in the literature – clearly, the traditional notion of human rationality must be reassessed in light of these findings.[45]

My own position is a defence of a pluralistic conception of human rationality. What the discrepancy between the actual responses given by participants in experiments with deductive reasoning tasks and the normative responses as defined by the traditional canons of deductive reasoning suggests is, or so it seems to me, neither that human reasoning abilities in general are faulty nor that the canons of deduction themselves are arbitrary, artificial constructions. Rather, one plausible conclusion to be drawn is that there may be different standards of rationality operating in different situations, all of which are presumably (at least to some degree) legitimate and adequate for the situations where they belong. The discrepancy emerges mainly as a result of applying a certain system of standards to the wrong situations. As the title of Counihan 2008a suggests, in many senses reasoning researchers have been 'looking for logic in all the wrong places'.[46]

---

[44] See Stenning and van Lambalgen 2008: 16 for the main positions in the debate.

[45] Similarly, these debates have sparked a reflection on the role of external, normative systems for research in psychology. In particular, Elqayam (Elqayam and Evans 2011) has articulated a rejection of what she describes as 'normativism' in psychological research.

[46] The view of rationality to be defended here is not a relativistic view: it is pluralistic in that it deems that there are different standards of rationality to be applied to different situations. However, the notion of normativity underpinning this claim is essentially an instrumental, pragmatist notion, taking as yardstick the goals to be attained.

As we have seen when discussing the experiments, it is far from obvious that participants univocally interpret what is expected of them in deductive reasoning tasks. As argued by many (in particular by Oaksford and Chater), once we consider the possibility that participants are in fact tackling a different reasoning task with the material given to them, many of the apparently deviating responses can actually be viewed as much more reasonable and adequate, as they are in fact responses to different tasks: the tasks *participants* think they should be solving, not the ones the *experimenter* wants them to solve.[47] But if this is correct (and to a large extent I think it is), then what these considerations also suggest is that *the game of deduction is not a game that we spontaneously play*, not in most cases anyway.[48] I cannot stress enough the importance of this idea for the present analysis; it is a mistake to look for deduction, in particular in connection with the concept of monotonicity, in humans' spontaneous reasoning patterns, grounded in their biologically determined cognitive apparatus. Rather, I will argue that deduction in fact corresponds to a method of reasoning and arguing developed specifically to be used in *scientific inquiry*; it is, in other words, a *cultural product* of the human historical development comparable to, for example, writing systems. And, just like writing, deductive reasoning (at least in a fairly narrow sense of the term, in particular concerning monotonicity and indefeasibility) typically requires specific training to be mastered.[49] This is supported by results reported by Evans, et al. (1994): elaborated instructions emphasizing the principle of logical necessity can significantly reduce (but not eliminate) the belief bias effect in syllogistic reasoning.

But, naturally, even if specific training is required, deductive reasoning is something that at least some of us humans *do* do, at least in some circumstances. So, obviously, it must be grounded in cognitive possibilities that are available in humans from the start. I here draw an analogy

[47] This still does not mean that the notion of *error* in such tasks is completely eliminated; participants may still approach incorrectly even the very task they think they are solving.
[48] This observation includes the specific, often contrived semantic interpretation typically given to terms of ordinary language in deductive settings. 'All', for example, is typically interpreted as an unrestricted quantifier, whereas everyday-life uses of 'all' are not (Counihan 2008b). Another well-known example is how, in ordinary language use, 'some' is usually thought to imply 'not all' (a so-called scalar implicature), whereas the 'logical' interpretation of 'some' has it that 'some A is B' is compatible with 'all A is B'. See also section 3.3.1, above.
[49] It is important to keep in mind though that in virtually every task there is a (varying) number of participants who do give the deductively correct response. Even in the original selection task experiment, 5 per cent of participants turned the 'right' cards. This suggests that claims of the total artificiality of the deductive canons with respect to human cognition are missing an important part of the story. Whether deductive competence only arises as a result of specific training and education, or whether it can arise even without specific training (in some cases), remains an important open question.

with Dehaene's analysis of the phenomenon of writing as a cultural product: it makes no sense to think that we must have an evolutionary 'writing module' (given our proficiency in writing), given that writing is such a recent development in human history. Rather, what seems to have taken place with writing is what Dehaene (2009: 303) calls 'neuronal recycling', 'writing's partial or total invasion of cortical areas that were originally devoted to a different function'. Learning to reason deductively may not lead to the same radical 'neuronal recycling', but I will propose that the origin of the deductive method in terms of more primitive capabilities is the human ability to engage in *dialogues* and to exchange different views and opinions with peers.

The view defended here is that the general patterns of non-monotonic reasoning that appear to underpin our spontaneous reasoning patterns, and which, according to Stenning and van Lambalgen (2008), are closely related to our general *planning* skills, are largely adequate and advantageous in most practical situations. However, in particular in the context of scientific inquiry, and possibly other contexts created by the needs and demands of modern life, our tendency towards doxastic conservativeness is not entirely advantageous. Hence, one possible account of the role of scientific method-ology is that scientific methods must provide mechanisms to compensate for our spontaneous reasoning tendencies when they are not entirely adequate in scientific contexts.

4.3.1 The usefulness of belief bias in everyday life

The term 'bias' still evokes negative connotations, so we may do well to reflect for a moment on whether doxastic conservativeness is indeed to be seen as a cognitive vice across the board. As already maintained, I believe it is by and large a very reasonable tendency when it comes to most everyday-life, practical situations, for a variety of reasons which I discuss now.

Going back to monotonic, deductive reasoning, an almost trivial but important observation is that deductive conclusions are 'costly' in the sense that a significant amount of information is required in order to produce just a few deductive conclusions. These conclusions do possess a high degree of certainty (assuming that we trust our deductive methods and our assumptions), but typically there are not many of them that can be drawn from given premises (often none). So, if deductive reasoning were to guide our practical reasoning, then, in most circumstances, decisions as to what should be done would simply not be forthcoming, as the available informa-tion often underdetermines the deductive conclusions that can be drawn.

Thus, the agent would not be able to act at all in such cases, which would most probably be quite detrimental to her well-being.

So, what do human agents do? In order to 'fill in the gaps' of the incomplete available information, humans call upon information in the form of prior beliefs, i.e., what they already assume to be the case. What else could they possibly do? It is evident that calling upon reliable prior information is the most rational thing to do here, especially if the alternative is not to be able to make practical decisions in virtue of insufficient information.

There are also computational reasons why doxastic conservativeness is advantageous. It is computationally cheaper to rely on a stable core of entrenched beliefs rather than to undertake belief revision at every single bit of new information. Moreover, if a hallmark of deductive (logical) reasoning is the inspection of *all* logical possibilities where the premises are the case, then clearly this is a computationally very costly procedure. By contrast, inspecting the much smaller class of preferred models requires fewer cognitive resources. More generally, given that humans are resource-bounded agents, it would in fact be highly inefficient in terms of allocation of cognitive resources (for practical purposes) to consider possibilities that are highly unlikely to be the case.[50]

Furthermore, purely in terms of practical rationality, it would seem that to err on the side of caution often offers a better pay-off than to be overly adventurous. If a particular sound *sounds* like a lion roar, even if the agent is not entirely sure, the safer bet is to assume that it *is* a lion roar and thus to run away as fast as possible. In more modern terms, if one happens to be in a subway wagon late at night together with a group of 'suspicious' individuals, it is again a safer bet (albeit perhaps politically incorrect) to trust one's own stereotypes and move to a different wagon.

Indeed, the idea that our 'fast and frugal' reasoning heuristics tend to offer advantageous pay-offs has been developed in detail by Gigerenzer and collaborators. They have argued that the fast and frugal algorithms they designed to model these heuristic mechanisms 'matched or outperformed all competitors in inferential speed and accuracy' (Gigerenzer and Goldstein 1996), and especially models based on a traditional conception of reasoning and decision-making. The 'fundamental computational bias' described by Stanovich is essentially one of these fast and frugal heuristics.[51]

---

[50] See Stenning and van Lambalgen's (2008: 16) discussion on the computational advantages of closed world reasoning with respect to classical propositional logic.

[51] The attentive reader may remember that the notion of preferential consequence leads to an explosion of computational complexity, a fact that appears to be in tension with the idea of belief bias as a fast and frugal heuristics. But what is computationally complex is to *model formally* the very relation of

The results of Gigerenzer and collaborators are significant in that they compel us once again to reflect on the very notion of standards of rationality. From this perspective, to consider the tendency to activate prior belief when reasoning as a 'bias' appears to be conceptually perverse. Even those who still use the bias terminology are well aware of this tension. Stanovich (2003), for example, acknowledges that this 'computational bias' is in most cases advantageous, for example, in terms of the allocation of cognitive resources and time constraints, and that it makes sense to think of it as an adaptation which increased fitness in the human environment of evolutionary adaptedness, many thousands of years ago (essentially as a 'logic of planning', as suggested by Stenning and van Lambalgen). But he claims that there are a few but important situations in *modern life* where such computational biases must be overridden, as they would provide suboptimal reasoning leading to negative real-life consequences.

It seems to me that, although he may overstate the negative consequences of belief bias in specific situations, Stanovich's observations are on the right track. As a general cognitive pattern, our tendency towards doxastic conservativeness is normally beneficial; but in specific situations, in particular in scientific contexts, it must at times be compensated for if we aim at optimal reasoning (given certain goals) in these situations.

4.3.2 Overcoming belief bias in science

Indeed, what is somewhat paradoxical about belief bias is that it conflicts with the very *raison d'être* of the scientific enterprise generally speaking, i.e., to discover *novel* facts about reality, and often (though not always) to replace previously held beliefs by new, more accurate beliefs. In the natural sciences (as elsewhere), one very important virtue in a theory is that it should make (preferably testable) *non-trivial predictions*, i.e., predictions that are somewhat counterintuitive and thus could potentially expand our current state of knowledge.[52] Now, if science is to be successful in its quest for new knowledge, it is to be expected that scientific methodology would encompass mechanisms that can act as a counterbalance to belief bias; otherwise, the replacement of prior scientific beliefs by novel ones might be compromised. How does the individual scientist manage to overcome her own entrenched

---

preference as a partial order of possible models, whereas choosing one's preferred models seems to be a fairly straightforward endeavour for humans. Here we seem to encounter the limitations of current formal models to model actual patterns of human reasoning.

[52] Albert Einstein is reputed to have said: 'If at first the idea is not absurd, then there is no hope for it.' Quoted in D. MacHale, *Wisdom* (London: Prion, 2002).

beliefs in order to arrive at new theories with non-trivial predictions? In other words, how do individual scientists manage to counter their own belief bias?

Naturally, the description of this cognitive phenomenon is likely to sound familiar to anyone acquainted with the philosophy of science literature; after Kuhn, the tendency to seek confirmation for the doctrines at the core of 'normal science' is recognized as a pervasive sociological phenomenon. Moreover, it is also well known that experiments *confirming* a given hypothesis tend to be viewed as more significant than experiments *falsifying* it. But what the results on belief bias suggest is that this tendency is not only a sociological phenomenon; on the strictly individual level, people in general have a strong tendency towards seeking confirmation for the beliefs they already hold.

At first sight, it may seem that a 'naturalistic' approach to scientific methodology, focusing on humans' actual cognitive patterns, will be chiefly interested in the relations of *continuity* between human cognition and scientific methodology. But the cases of *mismatch* are potentially even more interesting, i.e., situations of (apparent) clash between our spontaneous cognitive tendencies and the precepts of scientific methodology.[53] Such clashes suggest the existence of *cultural developments* leading to precepts that offer a counterbalance to the effects of these spontaneous patterns which may not always be advantageous in scientific contexts. Naturally, scientific practices are conducted by individuals having prima facie the same biologically determined cognitive apparatus as other human beings; but it is equally obvious that sound scientific practices require special training, and thus that an important aspect of this training consists precisely in learning to suppress more spontaneous cognitive tendencies.[54]

As we have seen, while in everyday-life situations we regularly make a large number of assumptions rather than relying only on the explicitly available information, there are a number of reasons why this is, generally speaking, not always good scientific practice. As clearly spelled out in the classical model of science presented in the *Posterior Analytics*, the goal is to

---

[53] Indeed, it seems that 'naturalized philosophers of science' typically adopt one of these two approaches, i.e., focus on the *continuity* (Gopnik and Meltzoff 1997; Carruthers 2002) or alternatively on the *discontinuity* (Dennett 1995) between spontaneous cognitive tendencies and reasoning in scientific contexts. Naturally, the latter also recognize that there has to be both continuity and discontinuity, so the difference is mostly one of focus.

[54] There is a rich literature on mechanisms of inhibition of biases, ranging from reasoning biases (Moutier, Angeard, and Houdé 2002; Houdé, et al. 2000) to racial biases (Amodio, et al. 2006). When discussing the debiasing effect of uses of formal languages, in Chapter 7, we will examine this literature in more detail.

start with (self-evident) truths and to progress with the highest possible degree of certainty towards new truths. Of course, in practice, scientific inquiry typically does not develop according to this model, but the general idea that scientific methodology should provide mechanisms aimed at preventing hidden assumptions from 'sneaking in' is crucial, precisely because systematically letting hidden assumptions 'sneak in' seems to be a pervasive feature of our spontaneous reasoning patterns. Interestingly, and while he was obviously unaware of the concept of belief bias as such, one of Frege's main motivations to develop a special notation for scientific inquiry was precisely to counter our tendency to let presuppositions 'sneak in'.[55]

One important reason why letting presuppositions sneak in unnoticed is not conducive to scientific practice is presumably that the degree of epistemic reliability of these presuppositions may be (and often is) inadequate. But, perhaps more importantly: if these assumptions are involved in the reasoning process, this process is much more likely to yield *confirmation* for the beliefs already held, as suggested by the concept of preferred models just discussed. Thereby, the discovery of new, surprising, perhaps counterintuitive facts is significantly hindered.

Thus, I essentially follow Frege in the idea that the use of a formal language is one of the means by which one can counter our cognitive tendency to let presuppositions sneak in unnoticed, i.e., to assume prior, external knowledge. Analysing in more detail why this is so in terms of the cognitive processes involved is the aim of the forthcoming chapters. But, more generally, the deductive method as such, even when practised without formal languages, can already been seen as offering a counterbalance to our belief bias tendencies, as I argue in the next section.

Naturally, the picture of scientific practices presented here is much too sketchy and brief; undoubtedly, it comes nowhere near accounting for the enormous diversity in scientific activity (Dunbar 2002). Moreover, it is somewhat in tension with the Kuhnian picture of normal science, given that I here stress the deterring role that doxastic conservativeness can have for scientific discovery.[56] To be sure, the Kuhnian model of normal science and scientific revolutions seems very plausible on a *descriptive* level; in effect, it both confirms and is confirmed by the prevalence of the phenomenon of belief bias, as described above. However, on a *prescriptive* level, to my mind

---

[55] Quoting a famous passage again: 'This deficiency led me to the idea of the present ideography. Its first purpose, therefore, is to provide us with the most reliable test of the validity of a chain of inferences and to point out every presupposition that tries to sneak in unnoticed, so that its origin can be investigated' (Frege 1879: 5–6).
[56] This is a reply to an objection put forward by Eros Carvalho.

Kuhn's model fails to emphasize the need to compensate for doxastic conservativeness for scientific discovery. The claim is not that doxastic conservativeness has absolutely no positive role to play in scientific practices; rather, the claim is that in scientific contexts it must be less preponderant than in everyday practices, which means that at least on some occasions it must be compensated for.

My thesis that logic and the deductive method in general, and logical formalisms in particular, can compensate for our tendency towards doxastic conservativeness has also been described as a form of 'Popperian Enlightenment'.[57] There is no doubt that I attribute a certain 'enlightening' effect to logic and logical formalisms; but I am well aware that scientific practices consist of much more than instances of hypothetical-deductive reasoning. The claim here is rather that *when* deduction is used within scientific practices one of the de facto effects it has is to compensate for doxastic conservativeness. In other words, I am offering a possible explanation for the cognitive strength of logic and formalisms in science, insofar as they are already being regularly used at different portions and different stages of the scientific enterprise; I am neither reducing scientific practices to the deductive method alone, nor recommending the deductive-logical approach as an escape from the darkness of the 'brute stages' of human cognition. The characterization of the present account of rationality as a *pluralist* account must be taken seriously.

4.3.3 *Deduction as a multi-agent, dialogical situation*

Let us take a step back and consider an apparently prosaic question: what would be a good antidote to making unwarranted assumptions? It seems to me that the answer is simple: *other people*, in particular those who disagree with these assumptions and/or are prepared to play the devil's advocate. Of course, like-minded people are not of much help in this respect, as their tacit endorsement of the same hidden assumptions is likely to reinforce our own existing beliefs.

Now, research on the historical development of the deductive method in ancient Greece – a unique event in human history – suggests that the social background of debates at the time played a crucial role in these developments (Lloyd 1996; Netz 1999; Marion and Castelnerac 2009). From this point of view, a deduction is primarily (originally) an *argument* (a discourse), not an inner mental process, more specifically an argument put

---

[57] In the terms of Paulo Faria, at the symposium on the manuscript in Porto Alegre on November 2011.

forward by a debater in order to compel other debaters to accept the conclusion of the argument if they accept its premises.[58] This idea can also be spelled out in game-theoretic terms: a deductive argument corresponds to a *winning strategy*; no matter what new information the other participants introduce (i.e., no matter what countermoves they make in the game), if they have granted the premises, they will be forced to grant the conclusion. Thus, *monotonicity* corresponds primarily to *indefeasibility*, and the multi-agent, dialogical situation would be the true conceptual *locus primus* of deduction. Deduction in mono-agent situations – both in linguistic situations (e.g., an agent conducting a mathematical demonstration) and in non-linguistic situations (e.g., an agent performing a mental process of reasoning) – is in fact a derivative notion.[59]

Naturally, I am not claiming that the Greek mathematicians and philosophers involved in the development of the deductive method had the explicit goal of offering a counterbalance to belief bias. Rather, insofar as they had a goal in mind at all, it was most likely that of formulating a style of debating that would constitute a superior alternative to other styles pervasive at the time – in particular the sophist style, geared towards convincing the audience at all costs (including by confusing and deceiving) rather than speaking truly (Netz 1999). But, in practice, and related to its dialogical origin, the deductive method even when deployed in mono-agent situations has a 'built-in virtual opponent', an opponent who, ideally, should contest the presuppositions sneaking in unnoticed and ask for appropriate justification at each inferential step. Each step must be completely watertight in that it must not allow for defeating counterexamples, no matter how far-fetched the counterexamples might be. In mathematics, the rationale of each inferential step must be entirely conspicuous and transparent so as to be fully justified.

Stenning draws similar connections between logic's historical origins and the idea of indefeasibility, and contrasts it with the nature of the majority of ordinary conversational situations which are typically cooperative:

---

[58] Mercier and Sperber (2011) defend an even more radical thesis: *all* reasoning is ultimately grounded in dialogical practices. 'Our hypothesis is that the function of reasoning is argumentative. It is to devise and evaluate arguments intended to persuade.' I do not endorse this radical thesis, as I think there is much reasoning that is non-linguistic and not geared towards linguistic expression (e.g., reasoning for planning). However, the thesis that deductive reasoning in particular has primarily an argumentative function is somewhat related to Mercier and Sperber's analysis.

[59] As I have argued (Dutilh Novaes 2011d), the idea that logic is primarily connected with *reasoning* rather than with *arguing* is a relatively recent idea, stemming essentially from Kant's transcendental idealism. Failing to recognize that deduction belongs primarily to public, multi-agent dialogical situations is one of the shortcomings of recent research in the psychology of reasoning.

Logic originated as a model of what we might call **adversarial** communication – at least in a technical sense of adversarial. What *follows* in deduction is anything that is true in *all* interpretations of the premises – that is the definition of logically valid inference . . . Our job, as speaker or hearer of a discourse that purports to be a deduction, is to test the drawing of inferences to destruction, to ensure that what is inferred is true in *all* models of the premises, not just the intended one. It is in this technical sense that logic models adversarial discourse. We may actually be socially co-operating in testing a deduction for validity, and certainly we have to co-operate a great deal to be sure that we are assuming the same things about the range of interpretations which are intended, but there is a core of seeking out all possible assignments of things, not simply finding one intended one. This is perhaps not accidentally related to the fact that logic arose as a model of legal and political debate. (Stenning 2002: 138)

Adopting this particular stance towards not only the origins but also the very nature of (classical) logic and deduction clarifies many of the puzzles that have been raised in this chapter. Recall the suggestion that the game of deduction is not a game that we spontaneously play. Now, while it is of course true that even very young children engage in adversarial forms of dialogical interaction (as any parent knows all too well), the particular form of adversarial dialogue underpinning the deductive method – the search for counterexamples no matter how improbable and far-fetched they may be – is arguably not part of our usual repertoire of dialogical interaction. While years of formal education may provide some familiarity with the idea of dialogues where one participant is 'testing' another (e.g., the teacher testing the student's knowledge), the exact form of the deductive approach to reasoning and arguing must (at least in the majority of cases) be *learned* upon intensive training (typically, by practice when learning how to formulate mathematical demonstrations). This learning process is some-what in tension with our more spontaneous forms of dialogical interaction, which are essentially about exchanging useful information and cooperating (as captured in the Gricean model of conversation).[60]

From this point of view, it is no wonder that participants who are untrained in the deductive method typically do not give the 'logical' responses in experiments with deductive tasks: they do not know the exact rules of the game of deduction, and instead tend to apply rules of other games that they are in fact familiar with (as also argued by, e.g., Oaksford and Chater). Viewed either as a mental process or as a form of

---

[60] Even that seems not to be entirely true. Recent studies (such as the work of Tomasello and collaborators) suggest that human verbal communication is much less about exchange of relevant information that thus far it has been assumed to be, and much more about social cohesion.

discourse, deduction contrasts with how humans typically reason and argue: seemingly, we reason following essentially non-monotonic patterns, and relying on something that could be described as 'a logic of planning'; moreover, we generally engage in dialogical practices that are for the most part cooperative.

But, interestingly, there are experimental results showing that, when solving deductive tasks in groups, participants' responses may converge towards the 'normative response' as dictated by the canons of deduction as traditionally construed. Moshman and Geil (1998) have applied the classical version of the Wason selection task to individuals and to groups of five or six individuals. While the 'correct' response pattern (as discussed in 4.1 above) was selected by only 9 per cent of the individual participants, 75 per cent of the groups selected this pattern. One may again conclude that deduction is a practice that is best engaged in collectively.

Some have claimed that, in spite of results that for the most part deviate from the canons of deduction as traditionally construed, participants do show a modicum of 'logical competence' (Evans 2002).[61] This was one of the original motivations for dual-process models of reasoning and cognition (Evans 2007; Frankish 2010), i.e., to account for this apparent discrepancy in the responses (mostly 'illogical', but occasionally 'logical').[62] Now, if there are indeed spontaneous, 'innate' mental processes that follow more or less closely the canons of deductive reasoning, and which participants engage in without specific training, it may well seem that this fact contradicts my claim that deductive reasoning is a cultural product (comparable to writing) which must be trained for to be mastered. I have two responses to this possible objection.

First, I have emphasized that, while it is a particular and somewhat contrived way of reasoning, deductive reasoning *is* related to some of our spontaneous reasoning patterns. For example, participants have no issues reasoning with something that looks like (classical) *modus ponens* at first sight but which is actually (as argued above) something quite different, namely reasoning with a defeasible conditional (not the material implication) and the assumption of background knowledge. So there are aspects of

---

[61] To be fair, the main point of Evans 2002 is in fact the declining role of logic and logicism in the psychology of reasoning; the focus of the paper is not the modicum of 'logical competence' that participants do display in experiments, but the latter is in fact mentioned.

[62] Currently, dual-process theories of reasoning are much more sophisticated than this brief description may seem to imply. Nevertheless, I believe there are general problems with the framework (e.g., Keren and Schul 2009), to be discussed more extensively in section 7.2.2, below.

deductive reasoning that resemble closely enough (while still differing from) more spontaneous reasoning patterns.

Secondly, and perhaps more importantly, one of the most serious methodological objections to the specific body of research that led many researchers to postulate the existence of two kinds of reasoning process concerns the demographics of the participants' samples. Almost invariably, research has been carried out with groups of undergraduate students at the researchers' universities, thus, typically, participants with many years of formal education of a specific, homogeneous kind (what may be inaptly described as 'Western-style schooling'). Now, if these participants display a 'modicum of logical competence', this is not entirely surprising given that they must have had *some* exposure to the main tenets of the deductive method throughout their school years, even if only implicitly. Thus, given the tendentious samples, defences of the *universality* of the dual nature of human cognition and the attribution of a 'logical' nature to one of the systems (System 2) on the basis of experimental results on such a limited sample of participants seem methodologically unwarranted.[63]

Even neurological results suggesting that different parts of the brain are involved when participants reason 'on the basis of belief' or 'on the basis of logic' (Goel and Dolan 2003; Goel 2007) still do not mean that these two styles of reasoning are both 'innate' in humans and both would arise without special training. As argued by Dehaene (2009), learning to read and write provokes a massive reorganization of the cortical structure in a human brain, and the vast majority of humans call upon roughly the same areas of the brain to learn to read and write. But, obviously, this does not mean that writing is not a cultural product; we do not have an 'innate' capacity to read and write. Similarly, years of training in 'school-like' reasoning may lead to a form of neuronal recycling similar to that of writing.[64] The work of Houdé et al. (2000; 2001) on the neural effects of cognitive inhibition training corroborates the claim that training can lead to the development of novel

---

[63] This methodological objection has been articulated in detail by Henrich, Heine, and Norenzayan (2010). From the abstract: 'Behavioral scientists routinely publish broad claims about human psychology and behavior in the world's top journals based on samples drawn entirely from Western, Educated, Industrialized, Rich, and Democratic (WEIRD) societies ... Overall, these empirical patterns suggest that we need to be less cavalier in addressing questions of *human* nature on the basis of data drawn from this particularly thin, and rather unusual, slice of humanity.' Their general point applies to the domain of reasoning as well: because it has been conducted almost exclusively with participants having a particular educational background, this body of research has not addressed the possible effects of education in reasoning.

[64] Indeed, it has been suggested (Goel and Dolan 2003) that 'logical' reasoning seems to activate areas of the brain normally involved in the processing of spatial information.

neural patterns.[65] Thus, none of the findings that purportedly support the dual-process model of reasoning seems to be outright incompatible with the view that so-called 'logical reasoning' is fundamentally a product of education, in particular given that very few comprehensive investigations have been carried out with participants having a different educational background, such as unschooled participants.[66]

This being said, the observation that a non-negligible proportion of participants do give the 'logically correct' response in reasoning tasks (ranging from 5 per cent in the original formulation of the selection task to 90 per cent with simple cases of modus ponens) is worth investigating in more detail. Somewhat against dual-process theories of reasoning, my hypothesis is that most of what could be described as 'deductive competence' is a product of training and education, as also suggested by other studies (e.g., Morris and Sloutsky 1998). In effect, I believe that what can be described as 'transformational-level' deductive competence (i.e., the reasoner actually drawing deductive inferences) is possibly entirely derivative or in any case greatly enhanced by training on the 'meta-level' of the very basic concepts of deductive validity and correctness. The results of Evans, et al. (1994) confirm that basic instruction on the notion of logical necessity seems to bring responses closer to the deductive canons, having a moderate debiasing effect. But, again, this is an empirical hypothesis which must be further put to test, as currently available results do not rule out the possibility that untrained participants, with no meta-level understanding of the concept of deduction, may nevertheless display competence on the transformational level.

At any rate, the main point is that while the deductive method was obviously not created specifically to counter belief bias it does seem to have this effect in that it features a 'built-in opponent' who questions the unjustified assumptions that the agent may make, forcing her to look at *all* the models of the premises rather than just at the much smaller class of her own preferred models.

4.4 CONCLUSION

My starting point in this chapter was the assumption that, to understand the effects of reasoning *with* formal languages, a general account of how

---

[65] The work of Houdé and collaborators will be discussed in section 7.2.1, below.

[66] (Luria 1976; Scribner 1997; Counihan 2008b; Counihan 2008a). As to be expected, the results of experiments with unschooled participants or participants with just a few years of formal education are very different from those with the typical undergraduate population.

humans reason *without* formal languages is required. So I have reviewed some of the findings from experimental psychology suggesting that human reasoning patterns deviate considerably from the canons of deduction and logic as traditionally construed. In particular, rather than only reasoning on the basis of explicitly available information (as the canons of deduction dictate), humans typically (and very sensibly!) reason taking background information and their prior beliefs into account. These features give rise to non-monotonic patterns of reasoning, which can be modelled in terms of the concepts of 'preferred models' and 'preferential consequence' borrowed from AI. Furthermore, I have argued that, while these spontaneous patterns of reasoning may be advantageous in most practical situations, they are somewhat in tension with the idea that the core of the scientific enterprise is to make non-trivial predictions and discover novel facts. They entail doxastic conservativeness, while in scientific contexts one must go beyond one's entrenched beliefs.

I have argued that the deductive method more generally offers a counterbalance to doxastic conservativeness, and this feature is related to the fact that it originated in adversarial dialogical situations, where one of the goals of each participant is to contest the others' assumptions. But, more importantly for the overall purpose of the book, in the coming chapters I will argue that *formal languages* are an even more powerful debiasing technology to counter doxastic conservativeness in specific contexts.

CHAPTER 5

Formal languages and extended cognition

The goal of this chapter is to formulate and defend some hypotheses regarding the cognitive processes involved in an agent's use of formal languages and formalisms when reasoning. The framework adopted will be that of *extended cognition*. It will be argued that rather than mere *expressions* of cognitive processes that take place independently, formal languages are *constitutive* of these very processes; a significant portion of reasoning with formal languages takes place literally 'on the paper'.[1] These claims apply to uses of formalisms more generally, including use of mathematical formalisms in physics, economics, mathematics itself, etc.

Given the almost trivial observation that formal languages are written languages, I start by reviewing some of the recent findings on the neuroscience of reading, essentially following Dehaene 2009, which may give us further clues as to what happens (and does *not* happen) on the neurological level when humans operate with different systems of writing. To my knowledge, there have been no such studies focusing particularly on uses of formal languages by human agents, but the available data allow for the formulation of some hypotheses.

I then venture into philosophy of mind to discuss the concept of *extended cognition*: the hypothesis that cognitive processes are marked by constant back-and-forth interactions between the agent and elements and items of her physical environment, thus entailing a focus on the perception-related ('bodily') components of these processes. Here, we will be interested specifically in the role of formal languages as constitutive of the very reasoning processes they are involved in. The main claim is that they do not merely express cognitive processes that take place independently; rather,

---

[1] That is, *whenever* formal languages are indeed involved. I certainly do not wish to claim that using formal languages is a necessary condition for a given practice to count as 'doing logic'. In particular, many of the investigations of logicians prior to the development of fully fledged formal languages (Aristotle, the Latin medieval logicians, etc.) should most definitely be seen as 'logic' even by modern standards.

they are inherently *part* of these processes, precisely because the latter appear to be intrinsically *modal* (as opposed to 'amodal', a terminology to be explained in due course). To illustrate the point, I discuss the results of Stenning and Landy and Goldstone (Stenning 2002; Landy and Goldstone 2007a; 2007b; 2009). These findings suggest that the physical, perceptual features of a given language do modify substantially the agent's reasoning processes when doing logic.

Finally, I develop the idea that formal languages do more than *extend* the mind:[2] they in fact *alter* the mind (albeit momentarily) in that they allow for very different reasoning processes to take place. They do so by eliciting a particular kind of bodily engagement from the agent, along the lines of what Landy and Goldstone (2009) have described as 'symbol-pushing'. Ultimately, this chapter elaborates on the idea that formal languages are formal in the sense of *computable* (as defined in Chapter 1) from a cognitive point of view, by arguing that strict observance of the rules of transformation of a formalism amounts to an *externalization* of reasoning processes, with significant engagement of sensorimotor processing.

5.1 READING IN GENERAL

The purpose of providing a brief overview of the neuroscience of reading in the present context is to discuss the sensorimotor aspects involved in reading and writing in general. Insofar as formal languages are written languages, and given the extended cognition perspective adopted here, it is imperative to go back to the basics of how human beings seem to process portions of writing. Section 5.1.1 focuses on the dual-route model of reading, which will also be crucial for the discussion of the concept of 'de-semantification' in the next chapter. Section 5.1.2 presents data suggesting that different kinds of writing seem to elicit different cognitive processes, presumably as a result of training, in spite of their superficial similarities. This concerns in particular the differences between reading/writing words and operating with digits. This discrepancy lends empirical support to Krämer's (2003) conception of 'operative writing'. Section 5.1.3 discusses the neural constraints in the choice of symbols, outlining the importance of an embodied/extended perspective which takes into account the constraints imposed by human cognitive characteristics on the development of cognitive artefacts and cultural technologies. Finally, section 5.1.4 focuses on the issue of the 'loop back' from the external realm into the brain, investigating the neural

---

[2] In the sense of the extended mind (EM) hypothesis (Clark and Chalmers 1998; Clark 2008).

transformations that seem to take place when we learn to operate with a new cultural technology.

In recent decades, research on the psychology and neurology of reading has made significant progress, to a great extent thanks to advances in neuro-imaging techniques; it is now possible to 'see'[3] with a fair degree of accuracy part of what goes on in the brain when a human reads. As a result, we now have a good number of robust hypotheses and a few confirmed facts on how the human brain processes written words. One of the most prominent researchers in the area is Stanislas Dehaene; he has also worked extensively on humans' numerical cognition, and provided an overview of work in the field in his book *The Number Sense* (1997). Recently, Dehaene undertook a similar enterprise with respect to the neuroscience of reading in his book *Reading in the Brain* (2009). In this section, I largely rely on Dehaene's account of the current state of the discipline.

A preliminary, almost trivial, and yet crucial observation is that, in terms of perception, written languages elicit primarily our *visual system*.[4] So an important aspect of the neuroscience of reading will concern basic visual processes and how the brain reacts to specific visual stimuli such as letters and written words. In this sense, the neural patterns involved will, at least in the first stages of cognitive processing, be fundamentally different from the patterns involved in speech recognition, and will be the same for different forms of writing.

Another important observation is that most of the data available concern *reading* exclusively, and the transposition to the neural processes involved in *writing* is by no means straightforward. For starters, while reading appeals essentially to the perceptual system of vision (as well as the oculomotor system), writing obviously also engages other motor skills, typically involving one's hands. Now, in the current state of technological development of neuro-imaging techniques, most of the experiments require the participant to remain immobile, which naturally makes the investigation of cognitive processes involving hand-related motor processes much more delicate. Moreover, while reading is essentially done 'in the same way', be it on a computer screen or a sheet of paper or what have you, writing with paper and pencil or typing on a computer are likely to elicit different cognitive processes. Thus, while in an analysis of uses of formal languages we are arguably more interested in what people do when they *write* formal

---

[3] But this 'seeing' must be taken with a grain of salt, as argued by Fine (2010).

[4] Of course, that holds of the human agents who are not visually impaired. Braille writing, for example, is a writing system which does not elicit vision.

languages than when they *read* formal languages, for now, given the currently available empirical data, considerations on the neuroscience of reading will have to do as our starting point.

5.1.1 *Two reading systems: the dual-route model of reading*

As we have seen in Chapter 2, according to a very widespread, 'common-sense' account of written languages, they are first and foremost the visual counterparts of spoken languages – the *phonographic* conception of writing. I have discussed some of the conceptual limitations of this account, but, as we will see now, there are also empirical reasons why it is an inadequate account of writing.

The phonographic account of writing predicts that when an agent is confronted with a portion of written language she would first of all convert it into (internal) speech and then process the outcome following the usual mechanisms for speech-processing. Interestingly, there were times in the history of humanity when reading in silence was by no means the norm (as suggested by the famous passage in Augustine's *Confessions* commenting on Ambrose's reading habits). Apparently, at such times, the usual way of reading was simply to recite the text out loud to oneself so as to produce understanding of its content. In such situations, indeed, the detour via speech was a necessary condition for understanding. But, of course, this is not how proficient readers currently read; at best, external verbalization is used as a preliminary step in the first stages of the learning process.

Nevertheless, one possible and prima facie not entirely implausible model of reading comprehension would be that, when an agent reads, inscriptions are silently mapped into sounds in the brain, and from that point onwards, understanding would proceed through the usual phonetic route, just as when a hearer listens to a speaker. This is known as the *indirect* model of reading, and it is indirect in that it postulates that comprehension proceeds by the mapping of inscriptions into sounds and then of sounds into meaning. The alternative account is the *direct* model of reading comprehension, according to which inscriptions are mapped straight into meaning without the detour via sounds. Which one is the most accurate model of reading is obviously an empirical question, as these two are prima facie both conceptually plausible.

As is so often the case in neuroscience, brain lesions provide a privileged vantage point to investigate the issue. It is well known that after the occurrence of a stroke or other forms of brain damage some patients have their reading abilities severely impaired, as a result of damage in the areas which were (presumably) involved in reading prior to the event. Different degrees of dyslexia can ensue from brain lesion. But, interestingly, dyslexia caused by brain lesion comes not only in different degrees but also in different *kinds*. In some patients, the ability to convert words into sounds is impaired, which means that they no longer manage to read infrequent words out loud like 'sextant', even when these words have a regular spelling. The same applies to words that are new to the patient, or invented words and neologisms. This is known as deep or phonological dyslexia (Dehaene 2009: 38). But these patients are still able to read frequent words (regardless of whether their spelling is regular or not). In fact, some of the reading mistakes they make are very revealing: they may read the word 'ham' as *meat*, for example, suggesting that the word 'ham' elicited some access to its general meaning but not to its sound.

The opposite situation has also been observed, namely patients who no longer seem to be able to access the meaning of a word directly; instead, they can only read by (silently) converting every written word into a sound, and then accessing its meaning via the meaning of the sound produced. This condition is known as surface dyslexia. These patients can read and understand words with regular spelling, but they are helpless when confronted with words with irregular spelling such as 'enough'. They resemble children who are just learning to read and who must reconstruct the sound of a word in order to be able to read it.

It is very revealing that patients with either one of these conditions are no longer proficient readers; this strongly suggests that both mechanisms (from inscriptions directly to meaning and from inscriptions to sounds and then on to meaning) are crucially involved in the reading of healthy, proficient readers. These observations have led researchers to formulate a dual-route model of reading, which postulates that there are (at least) two reading processes that operate in parallel. These pathways are partially redundant, but are both necessary for proficient reading, as shown by these two forms of dyslexia ensuing from brain lesion. In healthy readers, the two paths seem to be used simultaneously, depending on the nature of the word to be read. As summarized by Dehaene:

– Infrequently used words and neologisms move along a phonological route that converts letter strings into speech sounds.

– Frequently used words, and those whose spelling does not correspond to their pronunciation, are recognized via a mental lexicon that allows us to access their identity and meaning. (Dehaene 2009: 104)

The dual-route model of reading has been extensively investigated (Jobard, Crivello, and Tzourio-Mazoyer 2003), and the brain areas involved in each of the two routes have been to some extent identified by means of brain scans of healthy readers performing a variety of reading tasks. One such task is to let participants read invented words, for which no meaning can be retrieved, so that the areas activated are presumably those involved in the so-called indirect route (access to meaning via sounds). Another experiment consists of asking participants to perform different tasks with words, for example to identify words that rhyme (presumably, primarily eliciting an activation of the sound-route), or words that are synonymous (presumably, primarily eliciting an activation of the direct route).

Dehaene is careful to point out that although the dual-route model of reading is widely held to be correct in its rough lines, many of the details of the processes involved are still not fully understood. This holds both of the processes involved in the conversion from inscriptions to sounds and of the processes involved in the mapping of inscriptions to meaning. From a philosophical point of view, the paths connecting inscriptions to meaning are likely to be viewed as particularly relevant, so here are some of their key characteristics:

> Semantics mobilizes a widespread array of regions . . . Crucially, not one of them is exclusive to the written word. Rather, they all activate as soon as we think about concepts conveyed by spoken words or even images. (Dehaene 2009: 109)

What this means is that while the first stages of processing a written word are specific to the particular modality of writing, thus engaging specific visual perceptual processes (more on which below), after this stage is completed the processes related to meaning appear to be largely amodal in that they treat stimuli in the form of inscriptions, spoken words, or pictures as roughly equivalent. Thus, past the first stages of perceptually processing the stimulus, semantics appears to be largely an amodal affair.[5]

Notice, however, that the process by which the meaning of *complex* expressions is generated by a reader who reads sentences 'remains a frustrating issue'.[6] Naturally, as long as these processes are not properly

---

[5] A given cognitive process is usually described as *modal* when the perceptual mode of presentation of the stimuli (external vehicles) plays a decisive role in the process; if this is not the case, then the process is described as *amodal* (Barsalou 1999).

[6] Furthermore, 'The process that allows our neuronal networks to snap together and "make sense" remains utterly mysterious. We do know, however, that meaning cannot be confined to only a few brain regions and probably depends on vast arrays of neurons distributed throughout the cortex' (Dehaene 2009: 111). It is probably the complexity and diffuseness of these processes across the brain that makes them so elusive to the specific neural approach of this research tradition.

understood, our grasp of the meaning processes involved in reading will remain very incomplete.

Nevertheless, the discovery of two pathways operating essentially in parallel when the proficient, healthy reader engages with portions of written language is an important and somewhat surprising discovery. Although it to some extent confirms the phonographic conception of writing discussed in Chapter 2, for the most part it in fact suggests that reading is much more than converting inscriptions into sounds. In effect, it would seem that, in quantitative terms, the direct route is called upon much more often than the indirect route; the latter is reserved for unusual and/or unknown words. Whenever possible, readers seem to prefer to activate the direct route rather than the indirect route (which makes sense from a computational point of view, as a computationally simpler procedure).

What is the relevance of these findings for the project of investigating uses of formal languages and other symbolic formalisms? It would seem that an important conclusion to be drawn is that, even though formal languages are first and foremost *written* languages, as discussed in Chapters 2 and 3, the cognitive processes involved in uses of formal language will most likely be fundamentally different from the cognitive processes involved in reading and understanding 'ordinary' written language. True enough, such a conclusion could already be reached even without the neurological data (the phenomenological differences between these two different forms of writing are immediately apparent), but now it is also further substantiated empirically. Crucially, neither of the two routes involved in the reading of 'ordinary' written language can be called upon when 'reading' formal languages (not in a straightforward way, in any case). Clearly, the indirect route cannot be activated, given that formal languages are typically scarcely verbalized, and only in a derivative way. As for the direct route, we cannot rule out on conceptual grounds alone the possibility that formal languages *do* activate some sort of semantic connections of meaning, but, if they do, such connections are likely to be of a different sort (given that they will not depend on the meaning of individual words).

The issue concerns the extent to which formal languages perform a discursive, expressive function, as many seem to think (see Chapter 3). The neurological approach may allow us to investigate this issue also from an empirical point of view: it is not only a matter of whether formal languages can and should be viewed as expressive devices, on a conceptual level, but also of how a human agent *actually* processes bits of formal languages. Does she or does she not deploy cognitive processes which resemble the direct route of reading, going from inscriptions to meaning?

168 Extended cognition

This is by and large an empirical question, not a conceptual one, but, to my knowledge, neurological investigations with participants 'reading' formal languages have not yet been conducted. However, the differences in the processes involved in word-reading as opposed to (number) digit-reading, which have been quite extensively investigated by neuroscientists, allow us to formulate some hypotheses and predictions. So let us now turn to these findings.

5.1.2 Numbers and letters: different neural activations

Chapter 2 of Dehaene's *Reading in the Brain* begins with a discussion of a very famous case of brain lesion, a patient treated by J. J. Déjerine at the end of the nineteenth century; it was the first documented case of alexia resulting from brain damage. Following what seems to have been a stroke, Mr C, a well-read retired salesman, suddenly found himself unable to read. Importantly, the wide majority of his other cognitive functions remained intact, such as speech, face recognition, visual accuracy (for the most part), and, surprisingly, writing. But Mr C simply could not read the signs which he nevertheless recognized as letters: he could write but he could not even read the writing he produced.

Cases such as Mr C's are referred to as pure alexia, the complete inability to read. Presumably, the damaged areas of his brain were involved in the process of reading at its very early stages, i.e., before it 'bifurcates' into the two routes described above. But perhaps even more surprising than the fact that the patient could write but not read is that his ability to read and operate with Arabic digits remained largely unaffected. As summarized by Dehaene:

He read strings of Arabic digits with ease and even performed complex calculations without flinching. This essential observation implies that reading digits relies on anatomical pathways partially distinct from those used for reading letters and words. Visually speaking, however, digits and letters have very similar shapes that are arbitrary and interchangeable. In fact, a number of cultures represent numbers using letter shapes. (Dehaene 2009: 56)

That is, while there is nothing intrinsically dissimilar in letters and digits considered strictly as blueprints, once a human agent is trained in reading letters and operating with digits, apparently distinct neural patterns become established. (Similarly, while it is of course trivially true that 'lion' could have signified bread, once we learn to associate the word 'lion' with the animal, we can no longer force ourselves to think of bread at will instead of thinking of a lion upon hearing or reading the word 'lion').

Besides Déjerine's (1892) description of Mr C, a number of studies have described cases of patients who cannot read aloud letters and/or words but who can read aloud Arabic numerals (Albert et al. 1973; Hécaen and Kremin 1976; Anderson, Damasio, and Damasio 1990; Cipolotti, Butterworth, and Denes 1991). The converse has also been observed, i.e., patients with acalculia (the inability to read and operate with digits) but whose reading abilities remained largely intact (Cipolotti, Warrington, and Butterworth 1995). These cases of brain lesion illustrate almost beyond doubt that the ability to read letters and the ability to read and operate with digits activate different neural pathways, and thus are, at least to some extent, independent cognitive processes.[7]

The impact of these findings for an analysis of the cognitive processes involved in engaging with formal languages and other formalisms is significant. They imply that, while formal languages are *written* languages, the cognitive processes in question are likely to be distinct from the 'usual' processes of reading ordinary texts. In effect, based on the common *operative* nature of these writing systems (in the sense discussed in Chapters 2 and 3), I hypothesize that the neural patterns involved in processing formal languages may be similar to those related to dealing with Arabic digits. This hypothesis could in principle be tested empirically: if a trained logician were to suffer a form of brain lesion (which, of course, is not something we *hope* will happen) entailing some kind of reading impairment, it would be possible to investigate the dependence or independence of her different reading skills. If she is no longer able to operate with digits (being thus an acalculic patient), but is still able to operate with formal languages, this would attest to the independence of the relevant cognitive mechanisms, at least in this patient. If, however, acalculia is regularly accompanied by a loss of the ability to operate with formal languages (but not by general alexia), this would at least suggest that operating with digits and operating with formal languages activate closely related cognitive processes, which would differ from the processes associated with reading ordinary words. Again, the idea here is that an operative use of writing, i.e., a use of writing that is *constitutive* of certain cognitive processes (in particular calculative processes), underlies both our ability to calculate with Arabic digits following the algorithms learned in school, and at least some of the uses of formal languages and formalisms of trained specialists.

---

[7] In fact, there is evidence that operating with Arabic digits may also call upon different 'routes', just as reading letters. Cipolotti and Butterworth (1995) report on cases of impaired number transcoding and preserved calculation skill.

Of course, the odds of this constellation of circumstances occurring simultaneously (a professional logician suffering specific forms of brain damage so as to produce such effects) are very small, so we are not likely to obtain definitive answers to these questions by means of brain lesion studies. However, the well-documented fact that alexia and acalculia are conditions that can occur independently already suggests that different modes of reading and writing (and the difference is essentially *functional*, i.e., not related to perceptual properties of the symbols) can activate different cognitive patterns, presumably at least partially localized in different parts of the brain.

Three caveats seem to be in order, though. First, although an astonishing degree of homogeneity has been observed in patterns of neural activation for reading ordinary texts (both across individuals and across different languages and writing systems), we cannot be sure that the same degree of homogeneity would be observed specifically with respect to processing formal languages and formalisms. While letters and digits are typically learned by children at an early stage, and in fairly standardized ways, the same does not hold of formal languages, which are typically (if at all) learned only at later stages of cognitive ontogeny. At these stages, the brain is already organized in particular ways, a product of the person's previous experiences. Again, this is an empirical question, i.e., whether different trained specialists do or do not display the same patterns of neural activation when dealing with formalisms; my hypothesis is that there will be a considerable degree of inter-personal variation.

Second, it is often the case that portions of formal language are couched in texts of ordinary language, so the question could arise as to how different cognitive processes could be activated in alternation; how would the switch take place? But recall that even the reading of 'ordinary' texts is now thought to proceed by the parallel activation of (at least) two different routes, which are called upon depending on the nature of the word (mostly whether it is a frequent word or not). So there is no reason why a third pathway could not be operating in parallel, one dedicated to reading portions of formal language (provided, of course, that the person has received the relevant training).

Third, given that the results reported here concern almost exclusively *reading* but not *writing*, the differences in the neural pathways involved in reading words and letters and operating with Arabic digits do not rule out the possibility that *writing* words and letters is after all a cognitive process very similar to *writing* digits or portions of formal language. After all, Mr C could no longer read, but he could still write (besides being able to undertake calculations with digits). In other words, the hypothesis of operative writing as a cognitively different form of writing has yet to receive appropriate empirical confirmation.

At any rate, an upshot of the dissociation of alexia and acalculia is that there are good reasons not to assume that operating with formal languages is just a cognitive variant of reading portions of ordinary text. It may well be that fundamentally distinct cognitive processes take place, which in any case reinforces the characterization of formal languages as a form of *operative writing* (along with, e.g., the language of basic arithmetic or calculus) as opposed to other forms of writing.

5.1.3 Neural constraints on notational choices

One of the often mentioned characteristics of formal languages (as opposed to other forms of writing) is the alleged arbitrariness in the choice of symbols for a given language. The designer of a new formal language, or so the story goes, has absolute freedom in her choice of symbols; the act of imposition of denoting a given operation or entity by a novel symbol would be uncon-strained. In fact, given that most formalisms are developed to be treated as uninterpreted, this would again suggest that the choice of symbols is and should be unconstrained by semantic or semiotic considerations, that is if the symbols are to be amenable to different interpretations. As long as the syntax is well-defined and exhaustive, one given choice of individual symbols is as good as any other. Carnap defends precisely this point of view in *The Logical Syntax of Language*:

From the syntactical point of view it is irrelevant whether one of the two symbolical [isomorphic] languages makes use, let us say, of the sign '&', where the other uses '•' . . . In such cases the design (visual form, Gestalt) of the individual symbols is a matter of indifference. (Carnap 1934: 6)

As for the rules determining how the symbols are to be combined (the rules of formation of the language), it is of course not the case that any set of rules is as good as any other, even according to the received view. But, even in this respect, at first sight there seems to be a considerable degree of freedom, and different conventions for the concatenation of symbols have been adopted with no apparent significant difference for work in logic itself (e.g., the Polish conventions for scope v. the 'usual' conventions).

However, the purported arbitrariness in the development of a given formalism is somewhat in tension with the trivial observation that some notations are readily adopted by the scientific community while others are

not. A large number of notations fall into oblivion soon after their introduction. Why is it that some notations, in particular some specific formal
languages, are 'successful' in that they become widely adopted, while others
are not? If the choice of symbolism is entirely arbitrary, there should not be
such a significant variation in adherence.

Of course, nobody maintains that just any system of notation is as good
as any other. Some pragmatic considerations do play a role: the notation
must have a certain degree of simplicity so as to be easily learned and
manipulated; it should be easily reproducible in terms of typesetting and
other practical aspects; it should not be unnecessarily cumbersome.
Moreover, it would be naive to disregard the weight of historical tradition,
familiarity, and other sociological aspects; notations are inherently historical
objects, and, typically, innovations are well accepted only if they are made
necessary. Otherwise, the tendency is to continue using the body of
notations that a given scientific community is already familiar with.

Nevertheless, these considerations still do not seem to explain completely
why some notations are widely adopted while others are not. Some notations just 'work' better than others, and it is natural to consider the
possibility that the specific cognitive make-up of human beings imposes
important constraints on the kinds of symbols that such agents feel comfortable operating with. From this perspective, much of the historical
development of notations in general and of formal languages in particular
can be seen as a product of cultural evolution towards systems of notations
that are more manageable from the point of view of these specific agents and
their cognitive make-up.[8] Again, let me quote Landy and Goldstone's
insightful summary of the 'phylogeny' of mathematical notations:

Over the course of history, mathematical formalisms have evolved so that they are
cognitively helpful devices, and this evolution has entailed making apparently
superficial, but practically crucial, form changes. For example, the convention
introduced by Descartes (Cajori, 1927), in which letters near the beginning of the
alphabet are used to denote constants and those near the end to denote variables,
frees us from the burden of remembering which are which and allows us to use
our memory for other aspects of a mathematical situation. (Landy and Goldstone
2007a: 2039)

---

[8] We have already encountered the idea that the evolution of language towards highly compositional
signalling systems may be viewed as a form of cultural evolution (as investigated by S. Kirby and
collaborators). Dehaene (2009: chap. 4) develops similar ideas with respect to the evolution of writing,
as the search for systems that optimize the trade-off between expressivity and learnability, i.e., systems
that are better suited to our cognitive apparatus.

Indeed, formalisms are cognitive technologies which must be *usable* by agents with specific sensorimotor and cognitive characteristics but which must also be *useful* in that they allow these agents to perform cognitive tasks that would otherwise be more difficult or even impossible to perform.[9] Now, as suggested in Chapter 2, and still to be further developed in the coming chapters, I submit that one of the strengths of formal languages is precisely the debiasing effect that they may have. But, for this effect to obtain at all, specific formal languages must be *usable*, and must thus satisfy some basic constraints related to our own features as human beings.

First of all, given that most forms of writing amount to *visual* stimuli, obviously the ideal graphic patterns to be used are those which can be easily perceived and processed by humans. It is quite plausible that we have a particular stock of graphic patterns which form the building blocks of our visual acuity, and this is indeed what Dehaene (2009: chap. 3) maintains. He observes that all writing systems are based on a fairly limited set of basic shapes and patterns, which are then combined to give rise to different symbols. Moreover, neurological research on monkeys has shown that they are sensitive to the very same stock of basic patterns that seem to underpin our writing systems: certain neurons fire upon being shown a specific shape, others upon being shown another specific shape, and so on.

Perhaps the most striking feature of the inferior temporal neurons is that many of their preferred shapes closely resemble our letters, symbols, or elementary Chinese characters. Some neurons respond to two superimposed circles forming a figure 8, others react to the conjunction of two bars to form a T, and others prefer an asterisk, a circle, a J, a Y ... For this reason, I like to call them 'proto-letters'. That these shapes are so deeply embedded in the preferences of neurons in the brain of macaque monkeys is quite amazing. By what extraordinary coincidence is this cortical stock so similar to the alphabet that we inherited from the Hebrews, Greeks, and Romans? (Dehaene 2009: 137)

He then goes on to argue that these patterns are precisely those that are useful to distinguish the contours of visual scenes as they most often appear in our environment.

The most likely hypothesis is that these shapes were selected, either in the course of evolution or throughout the course of a lifetime of visual learning, precisely because they constitute a generic 'alphabet' of shapes that are essential to the parsing of the

---

[9] It is important that they should represent a significant improvement over cognitive processes unaided by such devices, otherwise the cost of learning how to operate with them would not justify their use; there must be a favourable trade-off between learning investment and pay-off.

visual scene. The shape T, for instance, is extremely frequent in natural scenes. (Dehaene 2009: 137)

For our purposes, it makes no difference whether this 'cortical alphabet' is innate or whether it is learned by every child upon exposure to essentially the same visual patterns as offered by a basic shared environment. What matters is that an important aspect of the learnability of a given writing system is that it should respect our neural preferences for certain shapes and patterns. We already have neurons that fire at the sight of particular shapes, so it makes sense that the development of a cultural technology should capitalize on this already available cognitive inclination. Nevertheless, exposition and learning also create new patterns of neural association between certain neurons and certain shapes or objects, especially at a young age when there is a great degree of cortical plasticity. 'By immersing children in an artificial environment of letters and words, we probably reorient many of their inferior temporal neurons to the optimal coding of writing' (Dehaene 2009: 141). In other words, writing adapts to the cognitive possibilities of humans, but individual humans must also adapt to the artificial shapes of writing. (This is probably yet another reason why it is typically difficult for an illiterate adult to become literate.)

From the point of view of the present analysis, these observations clearly indicate that even when it comes to the choice of individual symbols which are to compose a given notation/formalism such choices are also constrained by our natural preference towards certain visual patterns and shapes. Of course, the development of new formalisms tends to recruit shapes from the pool of already existing letters and symbols, but, even when it does not, newly created symbols are more likely to 'stick' if they echo the basic shapes of the 'proto-letters'.

But, of course, the choice of individual symbols is just the beginning of the story; what is peculiar about formal languages and formalisms is not only their individual symbols but, perhaps even more importantly, how they are combined. As noted before, a hallmark of a fully fledged formal language is the fact that it has an explicitly defined syntax, generating unambiguously its class of well-formed formulae. So we can ask ourselves what kind of sensori-motor anchoring the syntax of a given formal language has.

Now, with respect to individual formulae, rather than how formulae are articulated with each other, most formal languages make use of the left–right linear convention of many alphabetical writing systems. These formal languages clearly exploit the fact that many (most?) of us are trained in this particular reading convention. But notice that there is still room for different scope conventions, such as the already mentioned differences between the Polish notation and the 'usual' notations in logic.

However, more interesting than the linearity of individual formulae is how formulae are articulated with other formulas in proofs, inferential steps, and other transformational operations. I discuss this aspect in more detail below, but for now notice that for the notation to lend itself easily to these transformations certain structures on the formula-level will also be more advantageous than others. To anticipate some of the ideas to be developed later, the order of symbols in a given formula may have a facilitating effect for the application of the relevant operations. For example, take these two formally identical formulations of left-introduction of the implication in sequent calculus:

$$\frac{\Gamma, A \Rightarrow B, \Delta}{\Gamma \Rightarrow A {\rightarrow} B, \Delta} \qquad \frac{A, \Gamma \Rightarrow \Delta, B}{\Gamma \Rightarrow A {\rightarrow} B, \Delta}$$

The hypothesis to be developed later on is that the formulation on the left will have a facilitating effect for the application of the rule by a human agent because of the proximity of 'A' and 'B' in this formulation. It is as if the formula 'A' would simply 'glide' to the other side of the $\Rightarrow$ sign; this particular spatial configuration makes the relevant transformation to be undertaken more conspicuous. Hence, also within the formula-level, some configurations may be easier to operate with by a human agent in that certain familiar patterns of spatial transformation will mirror more closely the operation in question. Landy and Goldstone (2009) offer an analysis of the involvement of sensorimotor operations when manipulating formalisms, one which appears to substantiate these observations, and will be discussed in more detail in section 5.2.2.

5.1.4 *The transformative power of a new technology: reading and writing*

In Chapter 2, I suggested that one of the most remarkable features of technologies is that they often open up possibilities going well beyond the initial problems they were designed to address. Writing is the prototypical example of a deeply transforming technology: there is no doubt that mastery of this technology changes human cognition profoundly. This can be seen both in the historical development of different societies and in the individual development of a human being.

But what sort of transformation actually occurs, on the neural level, when a human being learns how to read and write? One of the most compelling ideas in Dehaene's book is that reading and writing skills depend crucially on what he dubs 'neuronal recycling' (a concept he also applies to other cultural skills, arithmetic in particular):

the 'neuronal recycling' hypothesis: the human capacity for cultural learning relies on a process of preempting or recycling preexisting brain circuitry. According to this ... view, the architecture of the human brain is limited and shares many traits with other nonhuman primates. It is laid down under tight genetic constraints, yet with a fringe of variability. I postulate that cultural acquisitions are only possible insofar as they fit within this fringe, by reconverting preexisting cerebral predispositions for another use. Accordingly, cultural plasticity is not unlimited, and all cultural inventions should be based on the preemption of preexisting evolutionary adaptations of the human brain. It thus becomes important to consider what may be the evolutionary precursors of reading and arithmetic. (Dehaene 2005: 134)

The idea of neuronal recycling is particularly appealing because it seems to be able to account both for the constraints imposed by the biologically determined cognitive make-up of humans and for the possibility of cultural variability and even of neuronal change as a result of the adoption of new cognitive technologies.[10] Specifically concerning reading and writing, Dehaene says:

We would not be able to read if our visual system did not spontaneously implement operations close to those indispensable for word recognition, and if it were not endowed with a small dose of plasticity that allows it to learn new shapes. During schooling, a part of this system rewires itself into a reasonably good device for invariant letter and word recognition. (Dehaene 2009: 146)

Of course, more than just the visual system is involved in reading and writing, but with respect to the other levels involved, one can also expect the pattern of neuronal recycling to be in place. We work with what we have, and, as already suggested in the previous section, the development of cultural technologies can be viewed as a process of cultural selection through which the technologies themselves gradually adapt to our cognitive possibilities. Obviously, we were not made to drive cars: rather, cars were made so that they could be driven by us. A similar principle arguably applies across the board to all cultural acquisitions.

An important aspect of the idea of neuronal recycling is the possibility of 'rewiring' related to the acquisition of new skills, which suggests that certain

---

[10] Even though, in this passage at least, Dehaene seems to play down the range of variability.

cultural technologies can indeed have a transformative effect on the brain itself.[11] So far, experimental work on the effect of literacy on cognition and reasoning has had inconclusive results; as it turns out, it is quite a challenge to isolate the effects of literacy properly speaking from the more global effects of schooling in general for experimental purposes.[12] On the neural level, it is again very difficult to isolate the effects of literacy while controlling for all other relevant factors.

However, in one study comparing the brains of pairs of sisters, one of whom could read and write while the other could not (Castro-Caldas, et al. 1998),[13] some small anatomic differences were identified between the two groups. 'Our results indicate that learning to read and write during childhood influences the functional organization of the adult human brain' (Castro-Caldas, et al. 1998: 1053). In a recent study, Dehaene, et al. (2010) also found evidence for the hypothesis that incorporating the use of a technology such as writing changes the cortical networks for vision and language. They discuss the following questions: 'Does literacy improve brain function? Does it also entail losses?' Their answer is 'yes' to both questions, even though there is stronger evidence for the former than for the latter.

Indeed, if the acquisition of certain cognitive skills prompts a significant change (even if it is not necessarily perceived on the neuronal level), it seems also legitimate to raise the question of possible cognitive *losses* entailed by the acquisition of new cognitive skills. If Dehaene is right, and neuronal recycling takes place, then presumably the parts of the brain now co-opted (rewired) for reading and writing may no longer perform their original functions (if any) in the same way as before. He raises this hypothesis himself, and suggests for, example, that learning to read and write forces us to abandon our natural tendency to 'mix left and right': 'we have to "unlearn" our spontaneous competence for mirror-image generalization' (Dehaene 2009: 293). In the Dehaene, et al. (2010) study, some modest evidence was found supporting the thesis that learning to read and write may interfere with face recognition.

More broadly, the idea that the acquisition of new cognitive skills may entail the loss of 'old' ones must be kept in mind concerning cognitive

---

[11] Menary (2010a) speaks of *cognitive transformations*.

[12] Counihan (2008a: chap. 1) provides an overview. Menary (2007a) presents a compelling account of the transformative power of writing from the point of view of cognitive integration. We will look into his analysis in more detail shortly.

[13] The assumption is that, with pairs of siblings, other relevant factors would be sufficiently similar, such as genetics, socio-cultural background, and environment.

technologies in general, and uses of formal languages in particular. Indeed, such technologies can, and apparently do, cause profound *cognitive transformations* (in Menary's fitting terminology); these transformations will probably be for the most part advantageous, but may also entail the loss of certain abilities (cf. section 5.2.3).

5.1.5 Conclusion

In this section, we have seen that some of the surprising findings in the neuroscience of reading – such as that there are (at least) two routes for reading, that the neural processes involved in digit-processing seem to be different from the neural processes involved in the reading of letters and words, and that we seem to have a preference for certain patterns of shapes when it comes to individual symbols – allow for the formulation of some hypotheses and conjectures concerning the cognitive processes involved in dealing with formal languages. In particular, they lend empirical support to the conception of formal languages as a form of *operative writing*, and as not having a primarily expressive function (as argued in Chapter 3). These considerations also highlight the fact that the notational choices for the development of a given formalism are constrained by very basic features of the human sensorimotor systems, which may explain why some notations are better suited for human use than others. In sum, reading/writing are clearly grounded in neural, cognitive, and sensorimotor possibilities that are present in humans from the start but which must then be co-opted (typically, upon extensive training) to be used for the different, novel task of operating with external symbolic systems – Dehaene's 'neuronal recycling' hypothesis. And operating with external symbolic systems (including 'ordinary' writing) is in fact a prototypical case of humans engaging in *extended cognition*, a concept to which we turn now.

5.2 FORMAL LANGUAGES AND EXTENDED COGNITION

In the last few decades, so-called embodied/situated/extended approaches to cognition have become pervasive in several fields, in particular in cognitive science and philosophy of mind. In the introduction to an influential article, Margaret Wilson thus describes the gist of the approach:

There is a movement afoot in cognitive science to grant the body a central role in shaping the mind. Proponents of embodied cognition take as their theoretical starting point not a mind working on abstract problems, but a body that requires a mind to make it function ... Traditionally, the various branches of cognitive science have viewed the mind as an abstract information processor, whose connections to the outside world were of little theoretical importance. Perceptual and motor systems, though reasonable objects of inquiry in their own right, were not considered relevant to understanding 'central' cognitive processes. Instead, they were thought to serve merely as peripheral input and output devices.[14] ... However, there is a radically different stance that also has roots in diverse branches of cognitive science. This stance has emphasized sensory and motor functions, as well as their importance for successful interaction with the environment ... This kind of approach has recently attained high visibility, under the banner of embodied cognition. There is a growing commitment to the idea that the mind must be understood in the context of its relationship to a physical body that interacts with the world. It is argued that we have evolved from creatures whose neural resources were devoted primarily to perceptual and motoric processing, and whose cognitive activity consisted largely of immediate, on-line interaction with the environment. Hence human cognition, rather than being centralized, abstract, and sharply distinct from peripheral input and output modules, may instead have deep roots in sensorimotor processing. (Wilson 2002: 625)

She immediately goes on to add that there are important dissimilarities among the different strands of embodied/extended approaches (see also below), but this would be the basic cluster of theses that all proponents seem to agree on.

Now, there is a sense in which an embodied approach to cognition is almost trivially correct: cognition is inherently tied to implementations in actual bodies. To this day, we have no record of cognitive processes having taken place in an entirely immaterial way (given the absence of evidence for the existence of immaterial thinking beings such as God or angels): a computer needs its hardware to perform its operations, and even Putnam's brain in a vat is, well, a brain in a vat! And, naturally, the brain in a vat is a thought experiment; nobody has ever encountered a 'thinking brain' detached from a functional body. Nevertheless, given the millennia-old pervasiveness of disembodied conceptions of cognition in philosophy and elsewhere – ranging from Aristotle's hylomorphic conception of the human being[15] to Cartesian dualism – it *is* important to stress the inherent connection between cognitive processes and the 'matter' in which they

[14] The fairness of this description of the bulk of research in psychology and cognitive science as being so thoroughly disembodied can be contested, but it is immaterial for the present purposes.
[15] To be sure, I suppose one *could* give Aristotle's account an embodied reading, insofar as he also stresses that, in the case of human beings at least, form can never occur by itself; it is always instantiated in matter.

occur.[16] In the case of human beings, this leads to a focus on the bodily aspects of cognitive processes, in particular concerning the engagement of sensorimotor functions and the relationship of the cognitive being with its physical environment.

A wide range of important philosophical debates take place against the general background of this approach to cognition; perhaps the liveliest of them concerns the *extended mind* (EM) *hypothesis* of Clark and Chalmers (1998), which has been extensively discussed, in particular since the publication of Menary's *Extended Mind* (2010b). This particular debate revolves chiefly around the specific (metaphysical) question of the boundaries of the mind, but it is important to realize that the adoption of an extended approach to cognition does not necessarily commit one to engaging in this particular debate (indeed, I will argue shortly that, for the present purposes, there are good reasons to *avoid* the debate altogether); these are independent issues.[17]

Moreover, notice that, while there are important commonalities, there are also significant differences between the concepts of embodied, situated, and extended cognition, and it is important to be clear on them. Here is an illuminating formulation of these differences:

> The literature makes a distinction between embodied, situated and extended cognition in supposedly ascending order of radicalness. The first claim says roughly that mind exists in the entire body, and not just in the central nervous system. The second claim says that certain environmental or social background conditions are necessary for certain cognitive functions. And the third claim holds that brain–body–world are dynamically coupled and thus mental states and cognitive functions might be viewed as extending spatiotemporally beyond the skin of the organism. (Chemero and Silberstein 2008: 129)

Thus, the embodied cognition perspective consists essentially in an emphasis on the role of the whole body in cognitive processes; the situated cognition perspective adds the relevance of the environment and in particular of social/cultural background conditions. Finally, the extended cognition perspective emphasizes the agent's interaction with non-biological realms in cognitive practices. It is mostly the last that underpins the debates on the boundaries of the mind and whether the concept of mind should be located exclusively in the brain, whether it should include other parts of the

---

[16] One need not go all the way towards physicalism, either, i.e., the view according to which all cognitive processes are ultimately to be reduced to material phenomena.

[17] 'EM, thus understood, is more an invitation to give detailed attention to [different kinds of interaction and couplings with the environment and its elements] in specific contexts and case studies than a fixed new metaphysics of mind' (Sutton 2010: 314).

body, or whether it extends even beyond the boundaries of the skin so as to include non-biological realms.[18] But, again, it need not be cast exclusively in terms of the boundaries debate; alternatively, it can be viewed as an approach to cognition which emphasizes the role of external 'things' (artefacts, media, or technologies) in cognitive processes (possibly also so as to include social and cultural phenomena, thus breaking away from an overly individualistic perspective).[19] The extended perspective will also naturally be concerned with the involvement of sensorimotor processing, as it is primarily by means of their sensorimotor systems that human beings interact with objects in the environment.

5.2.1 Engaging with external cognitive artefacts

As should be clear by now, the extended cognition approach is particularly suitable for the present purposes; we want to investigate the impact of using formal languages and formalisms more generally (conceived as cognitive technologies) on the relevant cognitive processes. The general claim will be that, rather than merely *expressing* these (purportedly independent) cognitive processes a posteriori, engagement with formal languages is *constitutive* of these processes.

But before we zoom into formal languages as cognitive artefacts, let us take a brief look at some of the examples of engaging with external cognitive artefacts in the literature. The classical example is Otto's famous notebook, as introduced by Clark and Chalmers. Otto has memory limitations, while Inga does not; they both want to attend an exhibition at MoMA in New York. Inga remembers that it is on 53rd Street while Otto does not. But Otto has the habit of carrying a notebook around. 'When he learns new information, he writes it down. When he needs some old information, he looks it up. For Otto, his notebook plays the role usually played by a biological memory' (Clark and Chalmers 1998: 16). So he looks up the address of MoMA and succeeds in his goal of attending the exhibition as does Inga; the difference is that he had to rely on the external object, the notebook, and the inscriptions in it.

---

[18] 'On the Extended Mind hypothesis (EM), many of our cognitive states and processes are hybrids, unevenly distributed across biological and non-biological realms . . . In certain circumstances, things – artifacts, media, or technologies – can have a cognitive life, with histories often as idiosyncratic as those of the embodied brains with which they couple . . . The realm of the mental can spread across the physical, social, and cultural environments as well as bodies and brains' (Sutton 2010: 289).

[19] Some recent critics, e.g., Hutchins, Kiverstein in presentations, have identified an excessively individualistic focus in Clark's conception of the EM, which is not organism-bound but remains organism-centred.

Ten years later, Chalmers begins the foreword to Clark's *Supersizing the Mind* with the following remarks:

A month ago, I bought an iPhone. The iPhone has already taken over some of the central functions of my brain. It has replaced part of my memory, storing phone numbers and addresses that I once would have taxed my brain with. It harbors my desires: I call up a memo with the names of my favorite dishes when I need to order at a local restaurant. I use it to calculate, when I need to figure out bills and tips . . . Friends joke that I should get the iPhone implanted into my brain. But if Andy Clark is right, all this would do is speed up the processing and free up my hands. The iPhone is part of my mind already. (Clark 2008: ix)

Despite the differences in technological sophistication, Otto's notebook and Chalmer's iPhone do not truly differ in terms of their participation in cognitive processes. Besides memory offloading, which is the archetypical use of external devices – shopping lists, notes-to-self, spatial distributions of objects such as the order of glasses for bartenders (Clark 2008: 62), etc. – such items can also be involved in performing other cognitive processes such as, for example, calculations (abacus, calculators, paper and pencil).

In a weak sense, the claim that external devices such as objects, media, and technology are involved in and facilitate cognitive processes is trivially true. No one would dispute that it is usually much easier to go to the supermarket with a well-crafted shopping list than to have to remember everything by heart. Importantly, though, the pay-off of the cost of involving such devices v. the benefits should be favourable: if you only need to buy two items, you will probably not bother making a list, which involves finding pen and paper, doing the writing, and making sure that you bring it along. But if the number of items you need is significantly larger (and the dividing line typically varies from person to person), then a shopping list is worth the trouble. So for the extended cognition approach (and, later on, the claim that formal languages transform cognitive processes in non-trivial ways) to have real 'bite' there must be more than incidental facilitation of cognitive processes which are otherwise essentially autonomous.[20]

Some of the 'bite' of the EM hypothesis comes precisely from the non-trivial proposal of redefining the boundaries of the concept of 'mind' on the basis of these observations. For this purpose, it is crucial to argue that external devices do not merely facilitate the cognitive tasks but are in fact *constitutive* of the cognitive processes in question (in a fairly strong sense of 'constitutive'). But, regardless of the boundaries issue (which calls upon the parity principle, to be rejected shortly), the idea of external artefacts being

[20] Cf. Menary 2010b: Introduction.

constitutive of cognitive processes is indeed crucial for my purposes. More specifically, I will argue that certain purely perceptual properties of specific notations in formal languages play a fundamental role for the cognitive processes in question.

More generally, this is not the place to argue extensively for the claim that artefacts and external devices are constitutive of a wide range of cognitive processes; corroborating evidence and arguments for this claim abound in the literature on extended cognition – Clark 2008 is an important source, but certainly not the only one; see also, for example, Kirsh 2010. But, to illustrate the point, let me recount the delightful anecdote that opens *Supersizing the Mind*:

Consider this famous exchange between the Nobel Prize-winning physicist Richard Feynman and the historian Charles Weiner. Weiner, encountering with a historian's glee a batch of Feynman's original notes and sketches, remarked that the materials represented 'a record of [Feynman's] day-to-day work.' But instead of simply acknowledging this historic value, Feynman reacted with unexpected sharpness:

'I actually did the work on the paper,' he said.

'Well,' Weiner said, 'the work was done in your head, but the record of it is still here.'

'No, it's not a *record*, not really. It's *working*. You have to work on paper and this is the paper. Okay?' (from Gleick 1993, 409)

Feynman's suggestion is, at the very least, that the loop into the external medium was integral to his intellectual activity (the 'working') itself. But I would like to go further and suggest that Feynman was actually *thinking* on the paper. (Clark 2008: xxv)

This is precisely the general idea to be pursued in the coming pages: 'thinking on the paper' with formal languages. Naturally, the external sensorimotor processes must latch on to relevant internal processes, so the claim is not that nothing goes on 'in the head' (or other internal parts of one's body). Rather, the claim is that *coordinated*, coupled internal and external processing characterizes these cognitive processes.

In this vein, Menary (2007a; 2007b; 2010a) speaks of *cognitive manipulations* as bodily engagements with elements of the environment which are often designed and created precisely to take part in cognitive processes. External representation systems, such as notations, formalisms, and formal languages in particular, are prototypical examples of especially designed elements of the environment – artefacts – that elicit bodily engagements and thereby play a role in cognitive processes. Think of the apparently trivial process of using paper and pencil to make a calculation:

Performing long multiplication involves mastery over a notational system. Humans very often create and manipulate external representational vehicles to complete a cognitive task. In doing so, they are carrying out a cognitive practice which is governed by its own norms – which I dub 'cognitive norms'. The norms are cognitive because they are aimed at completing cognitive tasks. (Menary 2010a: 565)

The fact that external graphic vehicles become available for manipulation is a general trait of any form of writing. As suggested by Menary (2007b), every form of writing has the operative dimension of allowing for a 'hands-on' implementation of cognitive processes.[21] To compose a philosophy paper, I typically write a first tentative draft, then review it, then move bits and pieces around (particularly with word processors), and the whole process is made possible by the very concreteness of the inscriptions.[22] But this is arguably even more the case when it comes to special notations such as mathematical notations and formal languages.

In the next section, we look more closely into the kind of cognitive manipulations that seem to be involved in operating with formal languages, and more generally with other notational systems. In particular, Menary's notion of 'cognitive norms' as mentioned in the passage above will prove to be very useful.

5.2.2 *Operating with formalisms as extended cognition*

The question to be addressed now is whether reasoning using formal languages and formalisms when reasoning is best understood as relying on amodal, internal structural rules and operation-templates (which are not straightforwardly related to explicit external perceptual characteristics of notations), or alternatively as depending crucially on (at least some of) the very physical, perceptual properties of notations.

For this end, I review the results reported by Stenning (2002) on the impact of using different languages for logic instruction, and some of Landy and Goldstone's results illustrating the significance of spatial, perceptual features of symbolic notation for the relevant reasoning processes. I take it

---

[21] 'If we take up the position of cognitive integration and see the external processes involved in writing working with neural processes, then we understand the entire co-ordinated result as an act of thinking. Further reasons in support of this explanatory framework were marshalled from the cognitive role of written vehicles and the resultant transformation of our cognitive capacities. These vehicles thus afford us new cognitive transformations which would be either impossible or extremely difficult by relying solely on neural resources' (Menary 2007b: 631).

[22] Notice, however, that I am certainly not claiming that this is the only approach to writing philosophy texts. Thomas Aquinas is reputed to have dictated different texts simultaneously to different scribes.

that these results strongly suggest that the agent's bodily engagement with the notation is much more significant for the reasoning processes involved than is often assumed.

So, what exactly is the role of notations when an agent is doing logic, mathematics, or any other intellectual activity that relies heavily on these devices? Prima facie, there seem to be (at least) two plausible but incompatible positions on the matter: (i) notations merely express (internal) cognitive processes that take place prior to and independently of their expression; (ii) notations have an active cognitive function to play in the very processes in question. The first account can be described as holding that '[s]ymbolic reasoning is proposed to depend on internal structural rules, which do not relate to explicit external forms' (Landy and Goldstone 2007b: 720). By contrast, according to the second account, notations actually *embody* the cognitive processes themselves, which take place through the 'paper-and-pencil' manipulation of the symbols. These two positions are represented in the relevant literature,[23] but it is fair to say that the first position, which stresses the irrelevance of (most) perceptual properties of notations for reasoning processes, is still more widely endorsed than the second one.

Here, I make two claims: (a) the second approach is a more accurate picture of actual reasoning processes and the use we make of notations; (b) it is precisely because it incites a form of perceptual, bodily engagement from the reasoner that reasoning with notations is a fundamentally different cognitive process, and thus offers an alternative to some 'default' modes of reasoning such as those related to the so-called belief bias phenomenon as described in Chapter 4. Therefore, reasoning with notations can, and in fact often does, counter our tendency towards relying on prior beliefs when reasoning (to be discussed in the next section and in the coming chapters). Interestingly, a hundred years ago Whitehead had already made a similar suggestion, in particular concerning the engagement of visual-motor processing when dealing with notations and the change in cognitive processes that it brings about:[24]

By relieving the brain of all unnecessary work, a good notation sets it free to concentrate on more advanced problems, and in effect increases the mental power of the race . . . [I]n mathematics, granted that we are giving any serious attention to mathematical ideas, the symbolism is invariably an immense simplification . . . [B]y the aid of symbolism, *we can make transitions in reasoning almost mechanically by the eye*, which otherwise would call into play the higher faculties of the brain. It is a

---

[23] These positions are surveyed by Landy and Goldstone (2007b: 720; 2009).
[24] Repeating a passage already quoted in Chapter 1.

profoundly erroneous truism ... that we should cultivate the habit of thinking of what we are doing. The precise opposite is the case. Civilization advances by extending the number of *important operations which we can perform without thinking about them.* (Whitehead 1911: 59–61)

In other words, by exploiting 'the eye', the cognitive processes in question are profoundly modified, in such a way that it could be said that one is not 'thinking' about what one is doing when proceeding in this way. It is precisely 'not thinking' about the operations that allows the reasoner to exclude insight and ingenuity, and thus to block the interference of 'sneaky' presuppositions.

Empirical evidence in support of this extended cognition approach to notations can be found in the work of Landy and Goldstone (2007a; 2007b; 2009) and Stenning (2002: esp. chap. 2). Stenning reports on a study comparing the effects of logic instruction using two kinds of formal languages: a traditional 'sentential' language, and an innovative diagrammatic language (Barwise and Etchemendy's 'Hyperproof' system). In particular, Stenning was interested in investigating the transfer of learning from logic to other kinds of reasoning tasks. Clearly, even though the two systems are formally equivalent, their perceptual properties are quite different. Thus, if learning logic was simply a matter of acquiring and developing the appropriate 'internal structural rules', one should expect that there would be no significant differences in results between the two approaches.

What Stenning's results show, however, is that students being exposed to each of the two languages improve their skills at different cognitive tasks. Before and after the course, students took a GRE (Graduate Record Exam), composed of two main kinds of tasks: those which are described as *verbal* tasks, and those which are described as *analytical* tasks. The idea was to measure the cross-discipline transfer of learning, i.e., if a logic course would improve the students' skills also in tasks that were not directly related to the material taught,[25] but also whether the two different formal languages being used for instruction (the traditional 'sentential' language and the diagrammatic language) would yield different effects in this respect. Not surprisingly, students taking the traditional course had on average a more dramatic improvement of their performance on *verbal* tasks, while students taking the Hyperproof course had a more dramatic improvement of their

---

[25] The results were overall positive: 'the good news is that the students substantially improved their scores on the GRE post-test relative to their scores on the pre-test in both the Hyperproof and the conventional syntactic courses. Logic teaching does improve performance on this test of general reasoning' (Stenning 2002: 66). But notice that there was an average *decline* in GRE verbal scores for those who took the Hyperproof course, as shown in Stenning 2002: 66 (Table 3.8).

performance on *analytical* tasks. More surprisingly, though, the results show that individual differences interact in interesting ways with exposure to different formal languages.

Besides the GRE tasks, before and after the course, students were also asked to solve so-called 'Blocks World' tasks, which are closely related to the framework of the Hyperproof system (cf. Stenning 2002: chap. 2). Hence, it was to be expected that the students taking the Hyperproof course would uniformly improve their skills in Blocks World tasks more significantly than the other students, but this is not what happened. Stenning provides a diagram (Stenning 2002: 67 (Fig. 3.9)) showing that the students who had scored above the median in analytical tasks prior to the course and then went on to take the traditional course *decreased* their posterior scores on Blocks World task, while the students who had scored below the median in the same tasks and took the same course actually *improved* their Blocks World scores. As for those who took the Hyperproof course, those with an above median GRE analytical task score *improved* their Blocks World score after the course, but those below the median had a *decline* in their Blocks World score.[26] This is particularly puzzling, as one would expect that *all* students would improve their Blocks World scores after taking the Hyperproof course.

In other words, different groups of students reacted differently to each of the methods, and it seems fair to conclude that at least part of this effect can be attributed to the different formal languages being used. Importantly, the instructor was the same (Jon Etchemendy himself), so presumably the main methodological difference was really the underlying language being used. Hence, it can be concluded that two notational systems with different perceptual properties have had a significantly dissimilar cognitive impact on the students' learning process and transfer of learning, thus lending support to the idea that learning logic is not just a matter of developing the appropriate 'internal structural rules'.[27] Moreover, the transformation of the students' cognitive abilities transferred across disciplines, as shown by the changes in scores (increase and decrease) in pre-tests and post-tests. Clearly, learning a new skill, and in particular learning a new technique to

---

[26] 'Those who did well on the constraint-satisfaction problems like office allocation improved their Blocks World scores when taught by Hyperproof. But those who did poorly on the constraint satisfaction subscale actually declined in their Blocks World score after taking the same course. What is more, the interaction reversed in the students who took the syntactic course' (Stenning 2002: 67).

[27] For the record, all students passed the course, so there is no doubt that they did learn logic, at least at the introductory level aimed at.

manipulate external representational systems, had a transformative effect on the students' cognition, and the effects of each of the two representational systems learned were quite different.[28]

Landy and Goldstone's work focuses on how the reasoner's engagement with the notation seems to exploit sensorimotor systems. In one study (Landy and Goldstone 2007b), they analyse the results of self-generated productions in the domains of handwritten arithmetic expressions and typewritten statements in a formal logic. In both cases, they claim to have 'found substantial evidence for spatial representational schemes even in these highly symbolic domains' (Landy and Goldstone 2007a: 2033), in particular concerning the systematic introduction of spacing having no logical/formal function to play. In one of the experiments, they asked participants to write out simple equations by hand. They remarked:

If, as we propose, formal notations automatically encode spatial relations corresponding to structural relations, then spacing in handwritten equations should reflect the formal structure of the equation. In particular, spacing around equality signs should be very large, since such signs denote, in all cases, the broadest partition of the sentence. (Landy and Goldstone 2007a: 2034).

So, since multiplication has priority over addition, for example, there would be increased proximity between elements of the equation being multiplied than between elements of the equation being added (naturally, participants had training in basic algebraic language). The results confirmed the initial hypothesis that spatial relations between individual symbols within the equation would follow closely the formal structure of the equation and how its different elements relate conceptually to one another, suggesting that participants used spacing to encode these relations. Similar results were observed with corpus analysis of expressions generated by participants interacting with a Web-based teaching tool for logic, this time concerning sentential, 'standard' logical languages.

In another study (Landy and Goldstone 2009), participants were asked to solve simple linear equations with one variable, displayed against a background that moved rightward or leftward. The idea was to probe the level of involvement of processes normally related to motion, and thus to investigate the extent to which symbol manipulation amounts to 'symbol pushing'. If there is a significant 'symbol pushing' component to it, then the direction of movement in the background would have an effect in

---

[28] The transformation of cognitive abilities caused by the acquisition of the capacity to manipulate new representation systems is also noted by Menary (2010a).

performance; if, however, this is not the case, then this exclusively perceptual element, namely the direction of movement of the background, should not influence performance in any way.

It was observed that solving an equation was facilitated when the background motion moved in the direction of the numeric transposition required to determine the unknown variable, thus suggesting the involvement of sensorimotor processing. It is as if the agent was 'moving' the very bit of notation itself to a different 'place' when solving the equation, rather than implementing the operation in an abstract way. Thus, Landy and Goldstone's work suggests that these transformations substantially engage perceptual and motor systems to be carried out: 'elements of the problem are "picked up" and "moved" across the equation line' (Landy and Goldstone 2009: 1). The agent literally 'moves' bits and pieces of the notation around to perform these transformations.[29]

Furthermore, the experiments with connectionist networks reported in Bechtel 1994 corroborate a similar conclusion, namely 'that ability in natural deduction itself may rely on pattern recognition abilities that enable us to operate on external symbols rather than encodings of rules that might be applied to internal representations'. The interest of Bechtel's results is to suggest that there are interesting connections between the account of operating with formalisms defended here and the framework of connectionism and neural networks, but which, for reasons of space, will not be further discussed here.

Taken together, these results strongly suggest that, rather than mere auxiliaries for the performance of internal cognitive processes based primarily on internal schemata and internal structural rules (as according to the 'Piagetian' discussed in Chapter 4), manipulations of notations in logic (and possibly also in mathematics), formal languages in particular, are *constitutive* of these very cognitive processes.[30] So it seems appropriate to speak of an

---

[29] Macbeth (forthcoming) argues that this is exactly how one should operate with the formal system presented in Frege's *Begriffsschrift*. However, strict rule-following and the engagement of perceptual-motor systems are two independent features; I am claiming here that most notational systems have both features, but one could claim that a given notation has one feature without having the other.

[30] Of course, this does not mean that one cannot 'do logic' without the explicit engagement of notations, but these will be quite different cognitive tasks, with potentially different results. I can here refer to the Latin medieval tradition in logic, which is highly sophisticated and yet does not rely on the technology of specific notations for its development. For example, upon reading Buridan's treatise on consequence (Hubien 1976), one cannot help being impressed by the level of logical complexity he attained while making use exclusively of the semi-regimented but rather cumbersome academic Latin employed at the time.

externalization of the reasoning process: when notations are involved, thinking also takes place on the paper (or whatever other medium), not only in the head.

Of course, it would be incorrect to attribute to the opponent of an embodied approach to the cognitive work done by notations the view that *all* perceptual properties of notations are irrelevant. Obviously, concatenation, which is a perceptual feature of notational systems, is crucial for any conception of the import of notations. Concatenation (both horizontal and vertical) presumably represents the order of operations undertaken, and plays a fundamental *formal* role in logical systems. However, some may still insist that even the role of concatenation is largely a conventional affair, and that any convention (e.g., 'standard' v. Polish notation) would be as good as any other, as long as it is implemented systematically. The point here is to argue that this is *not* the case: just as the choice of symbols is constrained by the general visual patterns that are already favoured by our visual system, arguably some concatenation conventions will be more aligned than others with the sensorimotor processing that appears to be called upon when we are dealing with notations.

As a final point, let me note that the results by Landy and Goldstone also seem to vindicate Krämer's (2003) claims concerning the inherent iconicity of any form of written language (see Chapter 2). The physical transformations corresponding to moving bits and pieces of the notation around go well beyond the (apparent) linearity of ordinary writing; when it comes to formal languages, the two-dimensionality of writing is particularly conspicuous.

5.2.3 *Formal languages: not only extending but* altering *the mind*

Thus, I have argued that formal languages are not merely secondary (albeit useful) devices for the cognitive processes involved when a logician is at work. Rather than being the a posteriori expression of internal cognitive processes, they are constitutive elements of the relevant processes themselves, in particular in that 'pushing symbols' is an important component. If this is true, then the specific perceptual properties of a given formal language or formalism are more significant than they are usually thought to be. I have also argued that the extended cognition framework, as developed within cognitive science and philosophy of mind, provides a privileged vantage point to develop this account of formal languages. However, the extended cognition framework comes in different versions, and some of the discrepancies among them are particularly relevant for the present purposes.

The most influential approach (still) to extended cognition is currently the one developed by Andy Clark, in works ranging from his seminal paper with Chalmers (Clark and Chalmers 1998) to his recent book *Supersizing the Mind* and his contributions to the recent volume on the topic edited by Menary (Menary 2010b). Clark's position on extended cognition is multi-layered and sophisticated, but it seems fair to say that one of its cornerstones is the parity principle: 'cognitive states and processes extend beyond the brain and into the (external) world when the relevant parts of the world function in the same way as do unquestionably cognitive processes in the head'[31] (Sutton 2010: 296)

Back to the seminal Otto–Inga example: to be able to argue that Otto's mind extends ('leaks') into his notebook, Clark has to emphasize the functional similarities between Otto's cognitive processes involving the notebook and Inga's, which happen entirely in the head. Since the functional outcome is the same (they both manage to go to MoMA), and Inga's recalling the address of the museum is indisputably a cognitive process, so should Otto's coupling with his notebook be viewed as a cognitive process. Hence, when discussing the metaphysical issues of the boundaries of the mind and arguing that they go well beyond skull and skin, Clark and others tend to emphasize the *similarities* between cognitive processes where external artefacts are involved and cognitive processes 'in the head'. Moreover, Otto's case may seem to suggest that external artefacts are particularly needed when a human agent's cognitive functions are malfunctioning in some way or another. But that is of course not what Clark wants to argue for: he insists that we are all 'cyborgs' from the start (Clark 2003), and that interacting with the environment is an intrinsic feature of our cognitive lives.

The parity principle has been forcefully criticized by opponents of the EM hypothesis, in particular Adams and Aizawa (2001; 2008).[32] The details of this critique need not detain us here, but the crucial point is that some critics seem to have taken a rejection of the parity principle as entailing a rejection of the extended cognition perspective as a whole. However, as argued by Sutton (2010), endorsement of the parity principle is in fact not inherent to the general approach, and there are many proponents of the extended cognition approach (perhaps even the majority) who at the very least do not take the parity principle to be as crucial.

In effect, Sutton (2010) describes two 'waves' in the EM framework: first-wave EM and second-wave EM. The main difference between the two is

---

[31] The original formulation is due to Clark and Chalmers (1998: 8).
[32] See also the Editor's Introduction (Menary 2010b).

that the former focuses on the parity principle and on the *similarities* between cognitive processes with or without external artefacts (mostly for the metaphysical reasons already mentioned) while the latter focuses precisely on the *dissimilarities* between these processes. Instead of the parity principle, second-wave EM is based on a complementarity principle:

[I]n extended cognitive systems, external states and processes need not mimic or replicate the formats, dynamics, or functions of inner states and processes. Rather, different components of the overall (enduring or temporary) system can play quite different roles and have different properties while coupling in collective and complementary contributions to flexible thinking and acting. (Sutton 2010)

Sutton relies on the distinction between 'exograms' (external symbols) and 'engrams' (the brain's memory traces), as proposed by Donald (1991; 2001), to note that what is characteristic of the EM perspective in general is the focus on our abilities to hook up with exograms. Second-wave EM in turn emphasizes the *differences* between processes involving exograms and processes not involving them, even if the functional outcome is roughly similar. Sutton then goes on to notice that it is important not to overstate the differences between exograms and engrams in that the boundaries are more fluid than one might think,[33] but these are useful concepts to proceed in the analysis.

Of course, those who emphasize the parity principle do not deny that exograms complement and to some extent modify cognitive processes in the sense that they, for example, offload memory demands, thus in many cases enhancing the agent's cognitive abilities. But, crucially, they typically argue that the same cognitive outcome could, at least in theory, be obtained without the involvement of exograms, such as in the Otto–Inga example.[34] As a consequence, the focus is on the increase in 'computational power' offered by the involvement of exograms rather than on how they might truly modify and transform cognitive processes (not just 'more', but something really different). This is a more 'high-tech' version of the old idea that the main impact of writing in cognition concerns the increase in working

---

[33] An example of this fluidity would be Clark's (2008: chap. 3) account of language, both oral and written – in particular, how being able to speak and the internalization of words (exograms migrating to the brain, so to speak) shape one's thinking 'in the head'. Words for numbers, in particular, seem to function in this way.
[34] Recall one of the views on motivations to adopt formal languages mentioned in 3.2.1, above: useful, but not indispensable.

memory, i.e., that external representational systems are no more than devices for storage of information (an 'external drive', as it were).[35]

A conceptualization of the extended cognition approach falling within Sutton's second-wave EM is offered by Menary (2007b; 2010a) in terms of the concept of cognitive integration:

[C]ognition is the coordination of bodily processes of the organism with salient features of the environment, often created or maintained by the organism. A coordinated process allows the organism to perform cognitive tasks that it otherwise would be unable to; or allows it to perform tasks in a way that is distinctively different and is an improvement upon the way that the organism performs those tasks via neural processes alone. (Menary 2010a: 563)

As must be clear by now, the present investigation fully embraces what Sutton describes as second-wave EM, and what Menary conceptualizes in terms of 'cognitive integration': we are interested in how the use of formal languages and formalisms deeply modifies and transforms reasoning processes. In a sense, the form of second-wave EM adopted here is a fairly radical one, even going beyond complementarity; one might say that the principle underpinning the present analysis is a transformation principle: exograms may, at least in certain circumstances, truly transform cognitive processes. In fact (and this is sometimes referred to as third-wave EM), as we have seen previously in this chapter concerning writing, a case can be made for transformations in the *cognitive agent herself* being brought about by the interaction with external cognitive artefacts. So it is not only the mind that leaks into the world; the world invades the mind as well. Menary refers to this general idea with the term 'cognitive transformations'.

In section 5.2.2, I have argued that, by means of notations and formal languages, the reasoning process is externalized, eliciting a form of bodily engagements with these peculiar kinds of exograms; such engagements are aptly described as 'cognitive manipulations' (Menary 2007a; 2010b). The externalization of the reasoning process in turn allows the reasoner to counter, or at least to mitigate, her own reasoning biases, in particular the ubiquitous tendency we seem to have to rely on prior belief and let presuppositions 'sneak in' (as described in Chapter 4). When sufficiently developed, the notation provides specific instructions on how to proceed (i.e., which 'moves' are allowed) by means of its well-defined syntax, and

---

[35] Menary (2007b) criticizes this view, arguing that the aspect of manipulation and transformation of the vehicles themselves is arguably even more significant than simply offloading memory demands. Interestingly, he relies heavily on the work of Roy Harris, whose conception of writing has been discussed in Chapter 2.

this is certainly the case in a fully developed formal language/formal system. In particular, in formalized deductive settings, an appropriate notation requires that all premises be made entirely explicit, and the transformational steps allowed for given premises are entirely determined by the rules of transformation within the system.

Thus, reasoning with formal languages typically (though perhaps not always) has two features that seem to contribute to the possibility of, when appropriate, countering certain patterns of cognitive processes taking place exclusively 'in the head'. One of them is the engagement of sensorimotor systems, providing a form of 'physical grounding' that differs from the grounding of, for example, conceptual metaphors (Lakoff and Nuñez 2000). The second feature is the fact that, in developed notational systems such as formal languages and formal systems, the rules of formation and transformation are explicitly and exactly formulated, thus constraining the moves available to the reasoner. She cannot let her mind wander 'at will', which would likely lead to some well-entrenched reasoning patterns (such as those described as 'reasoning biases'); she must instead strictly follow the instructions contained in the notation. In both cases, the externalization of reasoning processes plays a crucial role.

Here, Menary's notion of 'cognitive norms' is very useful. When discussing the concept of 'cognitive practices' and the role of cognitive manipulations, he says:

When we manipulate a public representation to complete a cognitive task, we are performing a cognitive practice. We are able to manipulate public representational vehicles because we acquire manipulative abilities that are governed by cognitive norms. Public vehicles, such as written languages and mathematical symbols, are tokens of representational systems. Such systems have their own norms governing how we are to manipulate token representational vehicles. These norms are cognitive as opposed to moral or social norms; they are directly tied to the completion of cognitive tasks. (Menary 2010a: 570)

He then goes on to introduce different kinds of cognitive norms, and the most relevant kind in the present context is what he dubs 'manipulative norms': 'These are norms for manipulating inscriptions of a representational system' (Menary 2010a: 571). Formal languages, formalisms, and other notational systems rely on explicitly and precisely formulated manipulative norms, which correspond to the rules of formation and transformation within the system. These are norms that have to be learned for each specific system, but what is perhaps the main challenge for the beginner is to learn how to operate 'formally' (in the sense of the formal as computable in

Chapter 1) with representational systems, i.e., to reason by strictly following manipulative norms.[36] Once this general skill is learned, learning to manipulate specific systems is usually not particularly difficult. What is more: once the technique is mastered, it may even be carried out purely mentally rather than relying on actual tokens of external representational systems, but this is the result of a process of *internalization* of a technique that is initially learned in the public sphere.

Relying on a Vygotskian perspective, Menary also offers a discussion of the inherently social character of most learning processes, which is made particularly conspicuous by the concept of the 'cognitive norms' to be learned (either mostly by exposure or by explicit training). He writes:

From a Vygotskian developmental point of view, higher cognition, for example reasoning and memory, appears first on the 'intermental' plane, in other words, in social interaction ... Cognition, then, is primarily a social phenomenon. (Menary 2010a: 568)

Naturally, Vygotski also recognizes that skills learned on the 'intermental' plane may then be internalized and used on the 'intramental', individual plane, but the starting point is a social one. This fits nicely with the process of learning logic and how to manipulate notations investigated by Stenning (2002); once the learning process is completed (based on the appropriate cognitive norms), manipulations of notations may be carried out internally. In fact, the public nature of external representational systems such as formal languages plays a crucial role in learning processes, but perhaps also in how logicians communicate with one another.[37]

But if formalisms and other systems of exograms can have such a transformative effect, can they not also give rise to what could be described as '*pathologies of extended cognition*'? Can the 'invasion' of the mind by portions of the world also alter the mind for the worse? This is most certainly a point deserving further scrutiny. We know that such processes of internalization can have deep and lasting effects; for example, once one has learned to count and calculate within the decimal system it becomes very difficult to think about numbers in other bases. We become fully accustomed to relying on

---

[36] Menary's concepts of cognitive norms and manipulation also make it patent that the formal as computable, as discussed in Chapter 1, does indeed have a deontic component. Think again of Turing's 'computer', diligently carrying out the instructions contained in the program of the machine.

[37] This last aspect will not receive much attention in the present context, but is also worth investigating in more detail.

the contingent properties of the decimal system for calculation (see De Cruz, Neth, and Schlimm 2010).

Now, especially in view of the phenomena of system imprisonment and system-generated problems discussed in section 3.3.3, it is perfectly plausible that the internalization of inadequate 'external tools' may lead to what could be described as *pathologies of extended cognition*, and to a change for the worse in an agent's cognitive capacities. Mostly for reasons of space, I will not pursue this line of investigation in the present context; however, this is clearly a theme that those working within the framework of extended cognition in general would do well to explore in more detail.

At any rate, one conclusion to be drawn at this point is that the general background adopted here is irresolutely that of second-wave EM. The main idea is that formal languages not only extend the mind: they in fact *alter* the mind.[38] By means of the (partial) externalization of reasoning processes, they allow these processes to run 'on a different software', as it were.[39] One of the 'built-in' softwares that we seem to come equipped with (described by Stanovich as a 'fundamental computational bias' – see Chapter 4) is the tendency to rely on external, prior belief when reasoning. A formal language/system offers a different 'software' to operate with, which is arguably implemented not (exclusively) by the internalization of its rules, as many seem to think, but to a great extent by the sensorimotor manipulations of the notation itself.[40]

5.3 CONCLUSION

The main goal of this chapter was to inquire into some of the basic cognitive processes involved in operating 'formally' (in the sense of formal as computable) with formal languages and formalisms. For this purpose, I reviewed some of the findings on the neuroscience of reading and drew some connections between these results and hypotheses, and the idea of operating

---

[38] The term 'altered mind' to describe this view of the effect of formal languages upon reasoning processes has been suggested to me by Andy Clark in a personal communication.

[39] But, notice that the use of the software metaphor should not be read as an endorsement of the computational theory of mind. It is above all intended as an illustrative metaphor. Moreover, the claim that these are different cognitive tasks should have important implications for the 'parity argument', which is central to the EM hypothesis. But, again, this is a topic for further research.

[40] Notice though that, upon training, a logician may in fact be able to 'internalize' the manner of operating with notations as external devices; but she would still be having a form of embodied engagement in that she would 'picture' the sensorimotor operations she would normally carry out with the notations as such. As noticed by Clark (2008: chap. 3) and Sutton (2010), skilled use of cognitive artifacts may in fact modify cognitive processes even when they take place exclusively 'in the head'.

specifically with formal languages. Once the ground had been prepared, I moved on to argue that a particularly fruitful approach to the role of formal languages in practices (of beginners or experts) is offered by the concept of extended cognition, in particular in its 'second-wave' formulation as presented by Sutton (2010). From this point of view, formal languages are naturally viewed as *cognitive artefacts*, as cultural products designed by humans to facilitate and enable certain cognitive processes (they essentially belong to the long history of development of mathematical notations, as described in Chapter 3). I have argued that, from an extended cognition point of view, manipulating formal languages and formalisms is constitutive of the cognitive processes in question; as such, the perceptual properties of a formalism become crucial in that these manipulations are bodily engagements which elicit sensorimotor systems in human agents. Moreover, I have argued that this externalization of reasoning processes does not only extend the mind, it actually *alters* the mind, albeit perhaps temporarily,[41] in that it allows the reasoning processes to run on a different software, as it were, countering some of the well-documented reasoning biases discussed in Chapter 4.

---

[41] I add this qualification because even skilled logicians who are proficient in using the technology of formal languages will typically resort to the default 'biased', non-monotonic patterns of reasoning in most practical situations. So the transformation is (fortunately!) not so deep as to modify their reasoning in ordinary life, generally speaking.

CHAPTER 6

De-semantification

In this chapter, we look more closely into formal languages as formal in the sense of *de-semantification*, a concept which has been mentioned several times before, but which has not yet been discussed in sufficient detail. Very roughly, de-semantification refers to a way of manipulating systems of writing which abstracts (partially or entirely) from the meaning of the expressions, and thus proceeds by strict application of the rules of transformation defined for the systems. The general claim defended in Chapter 5 was that these manipulations of writing systems significantly engage sensorimotor processing rather than occurring 'internally' in an amodal way, i.e., by 'internal' applications of the rules of transformations. Indeed, I claim that the whole process occurs primarily in the external realm.

However, one must bear in mind that emphasis on the cognitive role of de-semantification for reasoning with formal languages is also compatible with an internalist story, and also that de-semantification is not a necessary condition for the sensorimotor conceptualization of the cognitive import of formal languages either. For example, on Macbeth's interpretation of Frege's *Begriffsschrift* (Macbeth 2011), 'moving bits and pieces' of the notation is precisely how one should operate with Frege's ideography, and yet he clearly views it as a fully meaningful language (as is also patent in his famous controversy with Hilbert (Blanchette 2007), which will be discussed in more detail below). So these two aspects, de-semantification and sensorimotor manipulation of notations, are *conceptually independent* aspects of the cognitive impact of operating with formal languages; on the present account, however, they are closely related. In particular, I maintain that they both play a key role for the debiasing effect which is one of the main cognitive upshots of using formal languages and formalisms, to be discussed in Chapter 7.

We will take Krämer's (2003) conceptualization of de-semantification as our starting point. Soon enough, a number of interesting implications and connections will emerge, in particular with some of the crucial turning points in the history of logic since the end of the nineteenth century, and also with recent debates in neurolinguistics and the psychology of language and meaning. Indeed, de-semantification is a very rich and fruitful concept, which allows for a discussion of a number of apparently independent phenomena.

The chapter is structured as follows: in the first part, I start by discussing the concept of de-semantification from a philosophical and logical point of view. I then claim that de-semantification is not the only relevant phenomenon here, and introduce the closely related but different concept of *re-semantification*. I then move on to connect the concept of de-semantification to recent debates in psychology and linguistics; in particular, I discuss the hypothesis that semantic activation is an automatic cognitive process, which might (partly) explain why using meaningless symbols or made-up words with no meaning for the agent significantly alters the reasoning process. In the second part of the chapter, I present two case studies where de-semantification and re-semantification seem to have played an important role in countering the pull towards reinforcing prior beliefs and thus in allowing for the discovery of counterintuitive results: Maxwell's formulation of the theory of electromagnetism and Lewis's triviality results on the semantics of conditionals.

6.1 DE-SEMANTIFICATION AND RE-SEMANTIFICATION

6.1.1 De-semantification from a philosophical point of view

As already mentioned, Krämer (2003) introduces the term 'de-semantification' to refer to a wide range of related phenomena which up until then had no specific name to be referred to by. These developments are not exactly recent: as described in Chapter 3, much of the historical development of mathematical notations can be viewed as the search for an operative, de-semantified approach to symbols and the formulation of efficient notations for calculation.

As the very term suggests, the basic idea is of a process of removal of, or disregard for, the semantic dimension of written signs, so as to treat them as 'meaningless'. But what does it mean for signs to be meaningless? One natural response is to view de-semantified inscriptions as lacking a semiotic relation to anything external to them: they do not *stand for* anything (at least, not while being manipulated). Properly speaking, they are not *signs* anymore, and instead become self-contained objects. Krämer comments on the role of the sign for zero before there was consensus on whether it stood for a number on a par with the other (natural) numbers:

The rules of calculus apply exclusively to the syntactic shape of written signs, not to their meaning: thus one can calculate with the sign 'o' long before it has been decided if its object of reference, the zero, is a number, in other words, before an interpretation for the numeral 'o' – the cardinal number of empty sets – has been found that is mathematically consistent. (Krämer 2003: 532)

Thus, a characteristic of processes of de-semantification is a certain degree of 'metaphysical freedom': one is allowed to use and manipulate signs even if it is not clear whether they in fact stand for any existing 'thing'. In effect, as reported by Krämer (2003 n. 26), Leibniz remarked that (commenting on whether infinitely small or infinitely large numbers are 'actual' numbers in the context of the development of infinitesimal calculus): 'On n'a point besoin de faire dépendre l'analyse mathématique des controverses métaphysiques' (There is no need to let mathematical analysis depend on metaphysical controversies).

Besides metaphysical freedom, de-semantification also seems to entail a certain degree of what could be described as 'epistemic freedom'. As noted by Krämer, and discussed in Chapter 3,

signs can be manipulated without interpretation. This realm separates the knowledge of *how* to solve a problem from the knowledge of *why* this solution functions. (Krämer 2003: 532)

Of course, this separation of knowledge-how from knowledge-why may give rise to suspicions concerning surveyability and reliability – recall the need for epistemic justification of the notational techniques developed within the abacus tradition discussed in Heeffer 2007 and mentioned in Chapter 3. But, despite these legitimate concerns, if the notation somehow manages to establish itself as reliable, then its application typically represents a *cognitive boost* for the agent, precisely in the senses often discussed in the extended cognition literature. Moreover, notice that an effective and reliable calculating notation may also represent a *democratization* of knowledge: cognitive tasks which would otherwise only be carried out by experts can now be carried out by a wider range of agents.[1]

As discussed in Chapter 2, Krämer introduces another concept closely related to that of de-semantification: the concept of operative writing.

We can explicate this idea with a type of writing in which this 'process of de-semantification' is particularly apparent. We will name this process 'operative

---

[1] Commenting on Viète's notational innovations, Krämer writes: 'To be able to solve an equation is no longer a "secret art form" ("ars magna et occulta"), but rather becomes the form of knowledge of rules that can be taught and learned' (Krämer 2003: 533).

writing'. This modality of writing is commonly known, and misunderstood[2], as 'formal language' and represents one of the fundamental innovations in seventeenth century science. (Krämer 2003: 531)

The basic idea seems to be that operative writing concerns not only the (a posteriori) expression of thoughts and cognitive phenomena but also the very co-occurrence of certain cognitive processes which take place *through writing*. Krämer illustrates the concept of operative writing by means of the decimal place–value system and the accompanying algorithms for calculation developed throughout its history (see Chapter 3).

This system [the decimal place–value system] made it possible not only to depict all natural numbers with ten written signs, but also to calculate with numbers. The decimal place–value system is both a *medium* for representing numbers and a *tool* for operating with numbers. (Krämer 2003: 531)

The opposition seems to be between *medium* and *tool*: operative writing would have a tool-like dimension that other forms of writing do not have. However, if we take into account Menary's (2007b) compelling analysis of writing in general from the point of view of extended cognition and cognitive integration we are led to the conclusion that *all* forms of writing, insofar as writing is a prototypical *cognitive artefact*, do possess a tool-like dimension in that they are actively involved in (constitutive of) certain cognitive processes. So we might be led to the conclusion that *all* writing is operative writing. From the point of view of cognitive integration, the rejection of the view that a given system of writing merely expresses, a posteriori, metaphysically independent thoughts should apply to *all* forms of writing, not only to specific kinds such as the written language of calculus or the formal languages of logicians.

And yet, there really seems to be a difference between forms of writing where the semiotic import of the symbols involved plays a prominent role, and forms of writing where this is not the case. Alternatively, it may be suggested that the operative character belongs not so much to the form of writing as such but to how the agent in question approaches it, i.e., specific applications; in manipulating portions of writing, she might be guided predominantly by the semiotic relations of the symbols to something else (as I am, while writing these lines), or she might abstract from meaning and

---

[2] This is a puzzling observation: Why is using the concept of formal languages in such cases a misunderstanding? I can only hope to contribute to a firmer grounding of the concept of a formal language in the present investigation, in such a way that the association between operative writing and formal languages is no longer a misunderstanding.

manipulate the inscriptions based on other principles. One may use either approach, even with respect to one and the same system of writing, on different occasions. However, it is of course true that some systems of writing lend themselves more easily to the operative approach – in particular if the symbols do not evoke strong meaning connections, and, most importantly, if the syntax of the system is sufficiently developed so as to allow for 'mechanical' transformations.

We can thus see that operative writing is not only a tool for describing, but also a tool for cognizing, a technique for thinking that enhances intelligence. Long before the computer became a universal medium and a programmable machine, we developed the computer 'in ourselves', which is understood here as the cognitive use of algorithmic sign-languages that are *freed of the constraints of interpretation*. (Krämer 2003: 534; emphasis added)

In other words, while all writing systems allow for 'hands-on' cognitive manipulations (in Menary's terms), not all writing systems are able to perform this role when abstraction is made of their expressive, referential import, because in such cases there is no other level determining how to operate and manipulate them (as there is with writing systems with a strict set of rules of formation and transformation).

There is another point of Krämer's analysis which deserves further scrutiny. She seems to view de-semantification as a *necessary* condition for the cognitive boost effect (although her remark may also be read as a historical observation of how these developments actually took place):

the potency of these calculations is *always* connected to a move toward de-semantification: the meanings of signs become un-differentiated. (Krämer 2003: 532; emphasis added)

However, it would seem that the cognitive boost afforded by especially designed notations does not *require* the move towards de-semantification, as attested by formal languages such as Frege's 'ideography' which do not rely on this component. In fact, de-semantification by itself may even have the opposite effect, if nothing else takes the place of meaning-based guiding principles for reasoning (as in some of the nineteenth-century 'formalistic' arithmetic – see Chapter 1). The real source for the cognitive boost appears to be the combination of de-semantification with externalization of the reasoning process as described in Chapter 5, in terms of extended cognition. Hence, the view defended here is that de-semantification is neither a necessary nor sufficient condition for the cognitive boost effect, but it may greatly enhance it (for reasons which will become clear shortly).

As we have seen in Chapter 3, the development of mathematical notations is arguably marked by the search for a de-semantified approach to
symbols, and the formulation of efficient notations for calculation
(although this is obviously not the *only* driving force behind the development of notations). However, what we could describe as 'self-conscious
de-semantification' is a fairly recent phenomenon: it is not present in
Frege's *Begriffsschrift* at all, and, even with the early Hilbert of the
Foundations of Geometry, what seems to be going on is *re*-semantification
rather than *de*-semantification (see also below). I submit that the first self-
conscious, articulated defence of de-semantification with respect to formal
languages and formalisms is to be found in Carnap's *The Logical Syntax of
Language*. As discussed in Chapter 3, Carnap argues that a dissociation from
meaning when dealing with symbols is *required* for the sake of precision and
clarity and to avoid metaphysical tangles; he refers to this approach as a
formal approach (obviously in the sense of de-semantification):

A theory, a rule, a definition, or the like is to be called *formal* when no reference is
made in it either to the meaning of the symbols (for example, the words) or to the
sense of the expressions (e.g. the sentences), but simply and solely to the kinds and
order of the symbols from which the expressions are constructed. (Carnap 1934: 1)

Carnap then famously argued that 'logical syntax is the same thing as the
construction and manipulation of a calculus' (Carnap 1934: 5); in other
words, the recommended attitude towards a logical language is to treat it as a
calculus, with well-defined rules of formation and transformation. (Carnap
is careful to add that he is not claiming that every language is nothing more
than a calculus; rather, the point is that syntax – logic – only concerns itself
with 'the formal aspect of language'.) In particular, 'it will not be assumed
that such a symbol possesses a meaning, or that it designates anything' –
again, we have the absence of a semiotic relation with something external to
the symbol itself.

The general idea of treating logical formalisms as calculi, whose symbols
are devoid of meaning, features prominently in much of the twentieth-
century literature on the topic. The motivations given for such a move
towards de-semantification are usually related to increased precision and
exactness. On a more technical level, the adoption of a meta-level perspective is also related to viewing logical formalisms as mathematical objects
in themselves (*about* which one proves theorems), thus also encouraging
de-semantification (as discussed in Chapter 3).

In section 6.1.3, however, we will see that there are fundamental *cognitive*
implications in using symbolic systems devoid of 'meaning' when

reasoning, but which, to my knowledge, have not received the attention they deserve in the literature. But, before we move on, I introduce another concept, that of *re-semantification*, which covers closely related but dissimilar phenomena.

6.1.2 Re-semantification

If de-semantification is conceived as total disregard for meaning, i.e., as treating inscriptions as mere blueprints with no semiotic connection to anything external to themselves, then it seems fair to say that full-blown de-semantification is not an accurate concept to describe many relevant events in the history of logic, mathematics, and the exact sciences more generally. The Leibnizian/Carnapian ideal of treating all symbolic systems as calculi is to a great extent an idealization. Contrary to Carnap's (1934: Introduction) recommendation, even when formulating logical formalisms, arguably we typically let ourselves be guided by extrinsic semantic intuitions and considerations. Does this mean that the debiasing effect of de-semantification is unattainable or in any case not often attained?

In fact, it would seem that, besides de-semantification, a closely related process often takes place, perhaps more often than de-semantification itself: I will call it *re-semantification*. The idea behind re-semantification is that a formalism which is developed to characterize a specific phenomenon A can then be reinterpreted on another phenomenon B. While the formalism may bring along some of the presuppositions and intuitions that the agent has concerning phenomenon A, when applied to phenomenon B it may nevertheless have the effect of countering presuppositions and intuitions that the agent has *with respect to B*. Alternatively, if the goal is to study phenomenon A in a different setting, then reinterpreting it in terms of the objects belonging to B may strip away the incidental and non-essential properties of the objects belonging to A, thus also having a form of debiasing effect.

In the exact sciences, analogies are powerful heuristic tools; Nersessian (1984: sects. 4.1–4.2) speaks of 'physical analogies'. When such analogies entail the application of a given formalism to a phenomenon which was not the very starting point for its development, what we seem to have is precisely a process of re-semantification (an example is Maxwell's theory of electromagnetism, discussed below). In logic and mathematics, re-semantification also has a distinguished pedigree: Hilbert's technique of *reinterpretation* for (relative) proofs of consistency is an exemplary illustration.

Hilbert's consistency demonstrations in *F[oundations of]G[eometry]* are all demonstrations of *relative* consistency, which is to say that in each case the consistency of a set *AX* of geometric axioms is reduced to that of a familiar background theory *B*, demonstrating that *AX* is consistent if *B* is. The important technique Hilbert employs is the *reinterpretation* of the geometric terms appearing in *AX* in such a way that, as reinterpreted, the members of *AX* express theorems of *B*. For example, Hilbert's first consistency-proof interprets the terms 'point,' 'line,' and 'lies on' as standing respectively for a particular collection of ordered pairs of real numbers, for a collection of ratios of real numbers, and for an algebraically-defined relation between such pairs and ratios; under this reinterpretation, the geometric sentences in question express theorems of the background theory of real numbers. (Blanchette 2007: sect. 2)

Ultimately, such a process of re-semantification aims at de-emphasizing the specific properties of the objects/concepts referred to by the terms of a theory (for example, 'point' or 'line') so as to enhance focus on the logical relations between the concepts, i.e., the 'logical scaffolding' as defined by the axioms. In other words, re-semantification enhances emphasis on these logical relations as such, rather than on intuitions and presuppositions related to the very concepts/objects in question. With his technique of reinterpretation, Hilbert could counter or at least minimize the interference of the conceptual properties pertaining to points and lines as such. As he writes, in a letter to Frege:

[I]t is surely obvious that every theory is only a scaffolding or schema of concepts together with their necessary relations to one another, and that the basic elements can be thought of in any way one likes. If in speaking of my points I think of some system of things, e.g., the system: love, law, chimney-sweep . . . and then assume all my axioms as relations between these things, then my propositions, e.g., Pythagoras' theorem, are also valid for these things. In other words: any theory can always be applied to infinitely many systems of basic elements. (Letter of 29 Dec. 1899, quoted in Blanchette 2007)

The properties of the axiomatic system in question to be investigated (in particular satisfiability and consistency) are not properties related to specific characteristics of the concepts of *point* or *line*, but rather to what Hilbert describes as 'scaffolding or schema'. So, abstracting from the meaning of the geometric terms 'point' and 'line' allows him to adopt a favourable epistemic stance to investigate these matters. As pointed out by Blanchette (2007: sect. 2), these terms 'serve in Hilbert's work essentially as empty place-holders, susceptible of multiple interpretations'.[3] True enough, in later years, Hilbert moved more and more towards *de*-semantification, in particular in his meta-mathematics programme of the 1920s, but, regarding the early Hilbert of the *Foundations of Geometry*, the key concept is that of *re*-semantification, and one of the main motivations seemed to be to isolate specific properties of, and presuppositions related to, the mathematical objects under investigation.

The idea that formalisms are susceptible of multiple interpretations is of course at the core of much of the work done in logic and mathematics in the twentieth century, in particular with the advent of model theory. Shapiro (1998) draws a useful distinction between *algebraic* and *non-algebraic mathematical theories*. Roughly, non-algebraic theories are in principle about a unique model, and thus about a unique class of objects: the *intended* model of the theory (some examples would be arithmetic and mathematical analysis). Algebraic theories, in contrast, do not carry a prima facie claim to be about a unique model (examples are group theory and topology). Algebraic theories, insofar as they are about *hypothetical* classes of structures, which may or may not be instantiated, seem to realize Leibniz's ideal of freeing mathematics from metaphysical discussions; in that sense, they rely essentially on the idea of *de*-semantification. Non-algebraic theories, in contrast, are designed in view of a specific class of objects and their relations (the intended model), but may still be fruitfully reinterpreted on other structures (in some cases at least). In the latter case, in particular if the structures in question are entirely dissimilar, *re*-semantification seems to be the key concept.[4]

6.1.3 De-semantification from a cognitive point of view

Now, the semiotic relation between signs and what they stand for is not only of interest to philosophers. In effect, psychologists use the concept of 'semantic activation' to refer to the establishment of salient meaning connections by a speaker or reader (most studies focus on the case of reading) upon hearing or reading a word. In this section, I will argue that, if semantic activation is an automatic cognitive process (as evidenced by a plethora of

---

[3] As is well known, and aptly described by Blanchette (2007), Frege was very critical of this approach. To him, the connection between a theory and its target phenomenon (its 'intended model/objects') was crucial for its epistemic legitimacy, so any process of re-semantification or de-semantification was from his point of view hopelessly misguided and illegitimate.

[4] Similar considerations could be drawn concerning logical (as opposed to mathematical) formalisms.

empirical studies), a move towards de-semantification is a powerful way to 'turn off', as it were, some spontaneous reasoning patterns in humans, in particular the tendency to bring prior belief to bear when reasoning. Thus, the concept of de-semantification allows for a straightforward point of contact between philosophical and psychological debates.

To appreciate why semantic activation and de-semantification are relevant in the present context, recall the Sá, et al. experiment discussed in Chapter 4. When asked to evaluate the correctness of an argument with 'meaningful' words, participants seemed to rely predominantly on the plausibility of the conclusion on the basis of their prior beliefs.

> All living things need water.
> Roses need water.
> Roses are living things.

Only 32 per cent of the participants gave the logically 'correct' response when evaluating this syllogism, namely that it is invalid; arguably, this was because the conclusion expresses a widely endorsed belief. They were then given a little scenario of a different planet, involving an imaginary species, wampets, and an imaginary class, hudon, and subsequently were asked to evaluate the following syllogism:

> All animals of the hudon class are ferocious.
> Wampets are ferocious.
> Wampets are animals of the hudon class.

As we have seen in Chapter 4, 78 per cent of the very same participants whose great majority had failed to give the 'logically correct' response in the previous task gave the 'logically correct' response here, i.e., that the syllogism is invalid. Why is that? From a 'logicist' point of view, the two arguments are very similar (they share the same 'form'), but, from a more fine-grained cognitive perspective, they are widely dissimilar. In the first case, the terms used in the argument (and in the conclusion in particular) obviously produce a process of semantic activation, while in the second case this should in principle not happen, as participants have no meanings associations with the terms 'wampets' and 'hudons'. From this point of view, it is not particularly surprising that the majority of participants seem to deploy different reasoning strategies to evaluate the correctness of the arguments: in the first case, plausibility of the conclusion, and in the second case, some other strategy or strategies, which cannot be straightforwardly inferred from the experimental results alone. In any case, what is clear is that participants were not relying on the so-called 'logical form' of the arguments to evaluate their correctness, otherwise they would have been much more likely to converge on their assessments of the two arguments.

What would happen if participants were specifically instructed to disregard the meaning of 'living things', 'water', and 'rose' in the first case? Would they be more likely to judge this argument as invalid, as they did in the wampet/hudon case? In other words, can humans 'switch off' semantic activation at will? To address this question, let us look at some of the experimental results on semantic activation.

There is a wide range of results suggesting that semantic activation is largely an automatic cognitive process. This thesis has been contested in recent years, in particular in a paper by Stolz and Besner (1999) with the revealing title 'On the Myth of Automatic Semantic Activation in Reading', but proponents of the thesis have put forward a series of rejoinders (e.g., Heil, Rolke, and Pecchinenda 2004). The consensus now seems to be that, at least under a plausible characterization of what an automatic cognitive process is, semantic activation is largely automatic.

Let us review some of the data. One of the first observations of what could be described as the irresistible power of semantic activation is reported in one of the most widely cited papers in the history of experimental psychology, namely Stroop's (1935) 'Studies of Interference in Serial Verbal Reactions'. What is now known as the Stroop effect is the experimental observation that, when a colour word is written in an ink/colour other than the one named by the word (e.g., 'red' written in blue), naming the colour of the word (blue, in that case) takes longer and is more prone to error than when the colour of the ink matches the name of the colour. So it would seem that, upon reading 'red', semantic activation for the colour red occurs and trumps the retrieval of the word 'blue', which is the correct answer to the question, 'In what colour is this word written?' The Stroop effect is taken strongly to suggest that semantic activation is automatic, given that participants are not asked to read the word but simply to assess the colour of the letters, and yet it is as if they cannot help themselves and process the word semantically anyway. These observations have led to the formulation of the Stroop test, which is used in clinical practice and even in criminal investigations. In particular, it is used to uncover pretence illiteracy: if the Stroop test is applied, literates will invariably have a delay in response (while real illiterates will not) even if they are trying to pass as illiterates.

Other experimental results strongly suggesting that semantic activation is automatic are related to what is known as semantic priming effects. The seminal study is by Meyer and Schvaneveldt (1971), which presented facilitation effects in recognizing pairs of words if there is a semantic connection between them. The basic set-up for experiments on semantic priming involves pairs of words, one being the prime and the other being the target. Participants are exposed to the prime but not asked to perform any specific task (mere exposure); then comes the target, and they are asked to read the target word. It has been observed that, when the prime and the target are semantically related (e.g., 'dog' and 'wolf'), a facilitating effect takes place, i.e., the response time for the target word is shorter than in the absence of priming.[5] The priming effect has been investigated in a variety of sophisticated ways, and a range of qualifications and specifications has emerged, but the basic idea is captured in the description just given.

Another remarkable feature of priming effects is that they can occur even when the participant is exposed to a word for a very brief period of time, so brief in fact that the participant is not even aware that she has been exposed to a word (which appears in a flash, so to speak). This is known as *subliminal priming.*[6] Here again, the suggestion is that semantic activation has a strong automatic component, as priming occurs even when the participant is not aware of actually having 'read' a word.

Experiments which have put the automaticity of semantic activation into question (especially Stolz and Besner 1999) have shown that semantic activation is affected by spatial attention; under certain (suboptimal) spatial conditions of exhibition of the words, the typical effects associated with automaticity (Stroop and priming) are compromised. So now there seems to be agreement that certain spatial conditions are required for the process of semantic activation to kick in automatically, but this does not mean that semantic activation is not a largely automatic process (in particular under favourable conditions). Neely and Kahan conclude their survey article with the following remark:

> On the basis of the preponderance of evidence using this approach, we conclude that unless visual feature integration is impaired through misdirected spatial attention, [semantic activation] is indeed automatic in that it is unaffected by the intention for it to occur and by the amount and quality of the attentional resources allocated to it. (Neely and Kahan 2001: 89)

Let us now go back to the question formulated above: could participants voluntarily disregard the meanings of the words 'rose', 'water', and 'living

---

[5] Neely and Kahan 2001 provides an overview.
[6] 'One of the strongest pieces of evidence supporting an Automatic Spreading Activation account of priming is the subliminal priming effect' (Neely 1991: 297).

things' when evaluating the correctness of the first argument above, if asked to do so? If semantic activation is indeed automatic, it all seems to indicate that they could not, or only with much difficulty, even if explicitly told to do so.[7] The pull towards thinking of roses, water, and living things is very strong, and from that to making judgments on the basis of one's prior beliefs concerning these objects is arguably a fairly short step. By contrast, 'wampet' and 'hudon' are words that do not produce semantic activation in the participants (automatic or otherwise), so they could not rely on their prior beliefs concerning such 'entities' in order to judge the correctness of the argument.

In Chapter 4, I have argued that using made-up words such as 'wampet' and 'hudon' when formulating an argument is the closest one can get to the effect of de-semantification in participants who are not trained in using formalisms. But, upon adequate training (typically, in logic or mathematics courses), a reasoner can master the technique of dealing with formalisms in a de-semanticized way. The general point here is that using formal languages in a de-semanticized way (which is, as noted before, not a *necessary* condition to operate with formal languages) allows the reasoner to suppress, at least to some extent, the patterns of semantic activation and the closely related patterns of activation of prior belief. In other words, my claim is that the philosophical and logical concept of de-semantification also has the crucial cognitive effect of countering semantic activation, which then allows reasoners to deploy reasoning strategies which they would otherwise not rely upon spontaneously.

Semantic activation is also clearly closely related to one of the phenomena investigated within the psychology of reasoning tradition (discussed in Chapter 4), namely what is referred to as *matching bias*. Here is one proposed definition of the phenomenon:

The phenomenon known as matching bias consists of a tendency to see cases as relevant in logical reasoning tasks when the lexical content of a case matches that of a propositional rule, normally a conditional, which applies to that case. (Evans 1998: 45)

---

[7] Stolz and Besner (1999) argue for the weaker view that semantic activation is the default reaction of a reader upon reading a word: 'Our general claim is that processing words to the lexical/semantic level, although typically unintentional, is the default set rather than automatic in the sense of being inevitable, as widely claimed. This default set can be overridden in a number of ways.' Nevertheless, for the present purposes, the weaker 'default' view of semantic activation is still largely sufficient to support the claim that the move towards de-semantification is a powerful tool in terms of 'switching off' semantic activation, even if there may be other ways for this effect to come about.

In practice, it refers to the fact that, in Wason selection tasks, for example (see section 4.1.1), participants have a strong tendency to select the very items signified by the terms in the conditional rule they are told to verify. So, if the rule is 'If a card has an *even number* on one side, then it has a *vowel* on the other side' and the cards they are given are [A], [B], [4], and [7], one very common response is to turn [4] (even number) and [A] (vowel). This phenomenon has been observed in a wide variety of experimental set-ups, and resists even the introduction of a negating term (Evans 1998). So, if the conditional rule is 'If a card has an even number on one side, then it does not have a vowel on the other side', there is again a strong tendency among participants to select an even number and a vowel, and, as it happens, in this case, this *is* the 'correct response' according to the traditional deductive canons. However, it would seem that the pull towards matching bias is what motivates this response: 'correct' answer, but for the wrong reason.

In Chapter 7, we will look more closely into the results on matching bias when discussing the work on inhibition by Houdé and collaborators, but for now let me simply note that matching bias is clearly related to semantic activation: for humans, the default response when hearing/reading a word is to think of items that are signified by the word (in the example above, 'even number' and 'vowel'). It is somewhat surprising to notice that, to my knowledge, no mention is ever made of the semantic activation debate in the matching bias literature (or vice versa), while matching bias is clearly an instantiation of automatic, or default, semantic activation.

At any rate, when dealing with fully formulated formal languages, reasoners are advised to follow strictly the rules of transformation under-lying the system in question, which (as argued in Chapter 5) seem to rely substantially on sensorimotor processing. How de-semantification and the manipulation of bits and pieces of notation are cognitively combined to provoke a debiasing effect will be discussed in more detail in Chapter 7.

6.2 TWO CASE STUDIES

In this section, I present two case studies of how processes of de-semantification or re-semantification contributed to the formulation of surprising, non-trivial predictions about a given phenomenon, in particular in that such processes facilitated the suppression of firmly entrenched beliefs (even in those agents developing the theories themselves). They illustrate the claim that recourse to formalisms can act as a counterbalance to doxastic conservativeness.

6.2.1 *Maxwell and the theory of electromagnetism*

So let us first look into Maxwell's theory of electromagnetism.[8] For reasons of space, the discussion here will remain rather superficial, in particular disregarding some important aspects of the temporal development of Maxwell's ideas; for more detailed (but accessible) expositions, the reader is referred to the work of Stein (1970; 1981), Nersessian (1984: chap. 4), Laidler (1998: chap. 5), and Arianrhod (2003).[9]

Since the 1820s, it had been known that electricity and magnetism were closely related phenomena, but the exact nature of their relations remained unknown. A crucial step towards a better understanding of electromagnetic phenomena was Faraday's experimental discovery of 'lines of force'. In the period roughly ranging from 1831 to 1855, Faraday conducted hundreds of experiments in order to probe the nature of electromagnetic phenomena. In *Experimental Researches in Electricity* (published in three parts: 1839, 1844, and 1855), he reported on his findings, developing in particular the notion of a *field*, as, for example, formed by the lines of force observed around a magnet in experiments with iron filings (see Figure 1). Interestingly, there is not a single mathematical formula in his writings; Faraday was a self-taught physicist, a brilliant experimentalist but with no formal mathematical training. Indeed, Faraday discovered by experiment virtually all the laws and facts now known concerning electromagnetic induction.[10]

As is well known, Maxwell took Faraday's experiments as his starting point, in particular the idea of lines of force, as opposed to Newtonian action-at-a-distance, to study electromagnetic phenomena. Maxwell's initial goal was to give a mathematical formulation to the conception of lines of force; in doing so, he showed that there was an alternative to the action-at-a-distance conception of electromagnetic phenomena (as championed at the time by, e.g., Ampère). Thus stated, the endeavour may seem simple at first sight, i.e., something that any skilled mathematician could do with ease – not so. Expressing Faraday's ideas in

---

[8] This section has greatly benefited from discussions with Frederico Dutilh Novaes.

[9] Of course, these authors each offer different, and at times conflicting, interpretations of the development of Maxwell's theory. I do not claim that the interpretation I present here is the only suitable one; but it is a plausible and, I believe, essentially correct interpretation of the development of Maxwell's ideas. (Notice also that I do not claim to be making any novel contribution to the scholarship on Maxwell.)

[10] Faraday's case illustrates that the use of mathematical formalisms is not a *necessary* condition for discovery in science. Indeed, my main thesis here should not be understood in the sense of mathematical formalisms being the *only* way towards the discovery of novel facts; it is not even a sufficient condition. Rather, the suggestion here is that they *may* act as a counterbalance to doxastic conservativeness, when appropriate.

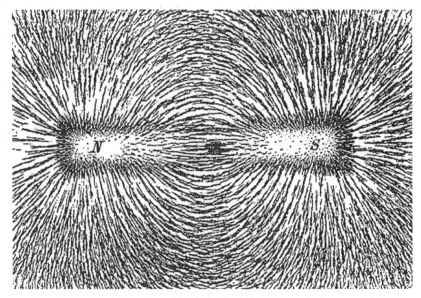

Figure 1 Magnetic lines of force around a bar magnet.
Source: N. H. Black and H. N. Davis, *Practical Physics for Secondary Schools: Fundamental Principles and Applications to Daily Life* (New York: Macmillan, 1913), p. 242 (Fig. 200).

mathematical form was anything but trivial; Maxwell also had to undertake a thorough conceptual analysis of these ideas. Nersessian (1984: sects. 4.1–4.2) argues that the method of 'physical analogy' played an important role in this process; the analogy was 'that between the intensity and direction of a line of force at a point and the flow of an incompressible fluid through a fine tube' (Nersessian 1984: 70). Having fluid flow in mind was a heuristic device allowing Maxwell to formulate a mathematical formalism which in fact could receive *different* physical interpretations, not only Faradayan lines of force. Notice, however, that the fluid analogy is firmly based on the idea that the 'mechanics' of electromagnetic phenomena requires a physical medium of propagation.

There are essentially three stages in the development of Maxwell's ideas, roughly corresponding to his three main papers on the topic: 'On Faraday's Lines of Force' (1855), 'On Physical Lines of Force' (1861), and 'A Dynamical Theory of the Electromagnetic Field' (1864). His mature theory is further developed in his complex book *Treatise on Electricity and Magnetism* (1873). The progression can be described somewhat paradoxically as characterized by an increasingly formalistic stance (in the sense that analogies and specific assumptions were gradually stripped off from the theory, which thereby became gradually more mathematical and more general in terms of possible physical interpretations), coupled with Maxwell's growing belief that, more than mere hypotheses, his theory might well be describing actual physical phenomena. This holds in particular of the conclusion that light *is* an electromagnetic wave – the theory predicts that the speed of propagation of an electromagnetic field is approximately that of the speed of light.

Importantly for our purposes, in the first two stages of these developments, the hypothesis of a physical ether as a necessary medium through which electromagnetic waves would propagate played a crucial role. Maxwell's true master stroke when formulating his final theory was to realize that his mathematical model could be made simpler and more general by the elimination of the ether hypothesis, or in other words by excluding the mathematical counterparts of his assumptions concerning the medium through which the waves travel.[11] But he *himself* still firmly believed in the existence of a physical ether, as clearly expressed in his article 'Ether' for the 1875 edition of the *Encyclopaedia Britannica*; see Laidler 1998: 181. So this is a remarkable example of a scientist suppressing his own tendency towards confirming prior belief.

By making his theory neutral with respect to the ether assumption, Maxwell would no longer reason on the basis of the traditional method of a visualized, mechanical model. The great William Thomson (Lord Kelvin) was among those who still held on to the traditional approach, and, indeed, he never completely understood Maxwell's theory (being an ether believer himself). But Maxwell let 'the mathematics speak for itself' (Arianrhod 2003: 263); he could thereby be neutral with respect to his belief in the existence of ether and attain conclusions (by following the syntax of the mathematical formalism) that were independent of the exact nature of the 'medium' through which the waves would propagate.[12] He treated the

---

[11] Here is Stein's (1970: 280) description of Maxwell's feat: 'Maxwell, therefore, in a truly fine piece of philosophical self-criticism, subjected his theory to an analysis, and showed that the essential contents of that theory ... could be preserved independently of any detailed account of the medium.'

[12] Here is a somewhat dramatic but interesting description of the role of mathematical formalism in Maxwell's theory: 'Hertz [remarked that] he could not escape the feeling that mathematical formulas have "an independent existence and intelligence of their own, that they are wiser ... even than their discoverers". In Maxwell's case, his equations were wiser than he was in that he still believed in the "probable" existence of a physical ether, even though his equations had made it all but superfluous' (Arianrhod 2003: 258). Notice that, as described by Stein (1970), Hertz played a fundamental role in clarifying the remaining loose ends in Maxwell's theory.

mathematical formalism as 'de-semanticized' rather than letting himself be guided by his own preferred interpretation. Now, while Maxwell's theory itself does not disprove the existence of a physical ether (it is mathematically compatible both with its existence and with its non-existence), by being neutral on the matter it paved the way for the elaboration of the theory which would eventually entirely disprove the existence of a physical ether, namely Einstein's theory of special relativity.

And, indeed, the theory made many surprising predictions, such as the existence of waves of different lengths (the electromagnetic spectrum), which made the discovery of radio waves possible; indeed, it had to be proposed as a theoretical possibility before experiments that could test it could be designed at all. More generally, Maxwell's theory of electromagnetism represents a historical turning point for what is to count as a physical theory; up to him, a physical theory was essentially a physical model, which could then be described mathematically. With Maxwell's mature theory, however, the mathematical formalism – the set of equations – *is* the theory, which can then be instantiated in different physical models.[13] This trend was further consolidated in the work of Lorentz, who explicitly rejected visualization as a reliable method in physics – see Nersessian 1984: chap. 5).

How exactly did reliance on his mathematical formalism allow Maxwell to resist the pull of doxastic conservativeness with respect to his belief in the existence of a physical ether? As I see it, two factors are crucial: the fact that the mathematical formalism allowed him to formulate a theory which was compatible both with the existence and with the non-existence of ether, and the possibility of letting 'the mathematics speak for itself', i.e., reasoning steps within the theory taking place on the basis of the mathematical syntax (in the spirit of de-semantification).[14] Both factors offer counterbalance to the tendency to resort to external prior belief, i.e., they allow for neutrality with respect to previous doxastic commitments.

Thus, we have seen that Maxwell started out with Faraday's 'picture' of lines of force, which he initially conceived of as tied to a specific physical medium, and then went on to elaborate a mathematical theory no longer based on visualization or mechanical models, and which did not make

---

[13] 'That Maxwell's theory is Maxwell's equations is even, in part, a biographical judgment: this is what Maxwell himself offered for us to believe' (Stein 1970: 282).

[14] Ironically, Maxwell was himself prone to making careless mistakes in his calculations, some of which even led to the 'right' results (see Nersessian 1984: 80–3).

assumptions as to the exact nature of such a medium. The possibility of expressing independence is indeed a trait of axiomatic approaches in general; axioms are (presumably) independent of each other, and hence dropping a given axiom renders a theory more general in that it applies both to situations where the dropped axiom does obtain and to those where it does not.[15] This is illustrated, for example, by so-called absolute geometries, which are compatible both with the acceptance and with the rejection of Euclid's fifth postulate. The crucial step in the development of Maxwell's theory was his success in eliminating the mathematical counterparts of the specific assumptions concerning a physical ether from the third formulation of the theory. So, in general, by isolating assumptions in the form of independent axioms, an axiomatic approach allows for the suppression of the belief in a particular assumption; it is sufficient that its counterpart within the theory no longer be viewed as one of its axioms.

Besides what we could describe as 'axiomatic freedom', insofar as Maxwell let 'the mathematics speak for itself' by following the rules of transformation of the calculus, the guiding principles for the reasoning process were no longer his 'intuitions' about the concepts in question (ether in particular). Instead, arguably we have a clear case of external-ization of the reasoning process, heavy reliance on the notation, and at least to some extent de-semantification or re-semantification. My main claim here is that these were the main features of his reasoning process which allowed him to remain neutral with respect to his belief in the existence of ether.

6.2.2 *Lewis on conditionals*

Conditionals are ubiquitous constructions in most if not all vernacular languages; and yet an appropriate semantic theory of conditionals has eluded generations of very able philosophers and semanticists.[16] That the logical material implication does not capture the semantics of conditionals in ordinary language is a view that has been widely (though not unan-imously) accepted for many years, but what should come in its stead is still

---

[15] See Schlimm 2011 on the creative role of axiomatics. His analysis focuses on mathematical examples – in particular, the development of the notion of lattice – but many of his general remarks apply to Maxwell's case too.

[16] For this section, I am greatly indebted to a presentation by, and conversations with, Hannes Leitgeb.

largely an open question.[17] Some theorists in fact hold that, unlike most kinds of statements in ordinary languages, conditionals do not have truth-conditions, properly speaking.

However, even if they do not have truth-conditions, it is still the case that there must be norms of assertability governing our uses of conditionals, given that there are clear differences in adequacy and assertability among different conditionals. Jackson (1991: 2) discusses 'If Jones lives in London, then he lives in Scotland' as an example of a highly unassertable conditional, but which on the material account (conditional as a material implication) comes out as true either if Jones does not live in London or if he lives in Scotland.[18] But now the burden is shifted elsewhere: even if we abandon the goal of formulating a truth-conditional account of the semantics of conditionals, we now need a robust account of assertion and assertability in order to spell out what it means for a conditional to be assertable/unassertable.

Much of the literature on conditionals of the past decades has thus focused on developing a theory of what it means for a conditional to be assertable (Adams; Edington; Lewis; Jackson). In particular, a significant portion of this literature adopted a probabilistic framework to discuss the semantics of conditionals, inspired by Ramsey's famous suggestion that conditionals concern *degrees of belief*:

If two people are arguing 'If p, will q?' and are both in doubt as to p, they are adding p hypothetically to their stock of knowledge, and arguing on that basis about q ... they are fixing their degrees of belief in q given p. (Ramsey 1929: 247)

The idea of adding p to one's stock of knowledge and then evaluating one's degree of confidence in q is known as the Ramsey test, and has been immensely influential in debates on the semantics of conditionals. The first to develop a fully fledged probabilistic account of conditionals was Ernest Adams (1965; 1975),[19] who defended the thesis that the assertability of a conditional is based on the conditional subjective probability of the consequent, given the antecedent. In other words, the probability of a conditional would be given by the conditional probability of the consequent given the antecedent, which is defined as:

---

[17] See section 4.1, above, on debates on the semantics of conditionals in the psychology of reasoning literature, and Edgington 2001 for an overview of the philosophical literature on conditionals.

[18] In this section we will focus on indicative conditionals, thus disregarding subjunctive conditionals, such as 'If Oswald had not killed Kennedy, then someone else would have.'

[19] Notice that, in psychology, a similar probabilistic turn took place only much later, but is now a very prominent approach to the investigation of conditionals (see section 4.1.1, above).

$$P(C|A) =_{df} P(CA)|P(A) \text{ (for } P(A) > o)$$

(P(CA) is the probability of both C and A occurring.) On Adams's account, it is permissible to assert an indicative conditional 'If A, then C' iff C is highly probable, given the occurrence of A. Now, one of the possible interpretations of this approach is the thesis that the conditional subjective probability (when defined, i.e., when P(A) > o) is in fact equal to the absolute probability of the conditional, P('If A, then C'). In a slogan: *'probabilities of conditionals are conditional probabilities'* (Lewis 1976: 298).

This general thesis was defended by a number of philosophers, Stalnaker (1970) in particular. In fact, there was almost unanimous agreement that the thesis was simply 'right', in all possible senses of the word; the semantics of conditionals had been nailed down. The account in terms of conditional probabilities clearly captured the main intuition of the Ramsey test, and was able to resist all the classical objections against alternative accounts (the one based on the material implication, in particular). Moreover, it did not seem to give rise to any obvious counter-examples, which is the prototypical way to question the cogency of a given theory according to traditional (analytic) philosophical methodology (e.g., Gettier cases, etc.).

But then came Lewis with his triviality results in 1976; Edgington (1995: 252) refers to this event as 'the bombshell'. Lewis (1976) showed that, departing from the thesis that the probability of a conditional corresponds to a conditional probability,

$$P(C|A) = P(A{\rightarrow}C)$$

and assuming very basic principles of probability calculus, one could derive an absurdity, namely

$$P(A{\rightarrow}C) = P(C)\text{[20]}$$

Now, obviously the probability of the conditional cannot simply be equal to the absolute probability of the consequent of the conditional! As in a typical *reductio* proof, deriving an absurdity entails that at least one of the elements used in the proof (either the premises or the inferential steps) would have to be revised/rejected. Lewis (1976: 304–5) does consider the

---

[20] For details of the proof, see Lewis 1976; Milne 2003.

possibility of rejecting one of the standard laws of probability used along the proof, but then notes that if one does that there is not much point in calling the probabilities in question 'probabilities' (i.e., if they do not comply with the standard principles of probability calculus). Thus, what is shown to be untenable by the proof is the thesis that the probability of a conditional corresponds to a conditional probability; otherwise, the conclusion would have to be that 'any language having a universal probability conditional is a trivial language' (Lewis 1976: 300).

Lewis's triviality results were interpreted as showing once and for all that the initially very plausible account of the subjective probability (and thus, assertibility) of a conditional in terms of the conditional probability of the consequent given the antecedent is untenable.[21] The key aspect for the present purposes is the fact that Lewis himself seemed to be favourably disposed towards this account ('this most pleasing explanation'), and that the 'formal', de-semantified application of standard transformation rules of probability calculus led him to the surprising results in question. Admittedly, such cases of 'de-semantified' reasoning leading to surprising results in philosophy are not as common as in physics and other exact sciences. But, Lewis's triviality results seem to offer a particularly suitable illustration of a non-trivial philosophical discovery obtained by the appropriate manipulation of a formalism (probability calculus, in this case).

6.3 CONCLUSION

This chapter focused on the concept of de-semantification from a number of different angles: philosophy, history, cognition. I introduced a closely related concept, that of re-semantification, which seems to be particularly pertinent for the analysis of some crucial episodes in the history of logic and science more generally (e.g., Hilbert's reinterpretation technique). It was also argued that, besides philosophical motivations (metaphysical and epistemic freedom), there is a distinctive *cognitive* basis for the epistemic boost afforded by processes of de-semantification. By countering our automatic (or default) tendency towards semantic activation, de-semantification allows for the deployment of reasoning strategies other than our default strategies, thus enhancing the 'mind-altering' effect of reasoning with formalisms (as argued in Chapter 5).

---

[21] 'The quest for a probability conditional is futile, and we must admit that assertability does not go by absolute probability in the case of indicative conditionals' (Lewis 1976: 298).

Finally, I offered two case studies to illustrate the debiasing effect of de-semantification: Maxwell's formulation of the theory of electromagnetism and Lewis's triviality results on the semantics of conditionals. So we now have all the pieces in place to move on to the final chapter of the book, where the debiasing effect of formalisms and formal languages is discussed in more detail.

CHAPTER 7

The debiasing effect of formalization

In Chapter 4, the concept of cognitive biases was presented, and several of the biases identified by research in psychology were discussed. We discussed in particular *belief bias*, and more generally what Stanovich has described as 'the fundamental cognitive bias in human cognition', namely the tendency to bring prior belief to bear when reasoning. I have argued that this is one of the main factors leading to a tendency towards doxastic conservativeness in human agents, i.e., the tendency to 'hold on' to the beliefs one already has.

Within the 'heuristics and biases' research programme initiated by Tversky and Kahneman in the 1970s, the question naturally arose as to what, if anything, could be done about the systematic deviations from certain theoretical standards in decision-making, judgment, and reasoning, as identified in the experiments. To refer to strategies that might counter-balance and correct these deviations, the concept of 'debiasing' was introduced:

That the mind has its illusions is not without dispute. Thirty years of decision research has used rational theories from economics, statistics, and logic to argue that descriptive behavior falls systematically short of normative ideals. But this apparent gap between the normative and the descriptive has provoked many debates: Is there in fact a gap? And, if there is, can it be closed – that is, can biases be 'debiased'? (Larrick 2004: 316)

The concept of debiasing has been widely studied in the literature on judgment and decision-making (Larrick 2004; Fischhoff 1982), but less so by researchers working specifically on reasoning (for exceptions, see section 7.2, below). Stanovich (1999; 2004: chap. 6) has coined the term 'Meliorists' to refer to those (such as himself) who believe that human agents can be made to perform much closer to theoretical/normative standards upon instruction and training, i.e., that debiasing *is* possible. Others (the 'Panglossians' in Stanovich's terminology) believe that there is nothing to be 'debiased' – there is no real gap between the normative and the descriptive; and yet others (the 'Apologists') maintain that biases are so deeply engrained in our cognition that no effective, long-term debiasing measures can be found – people are reasoning about as well as they possibly can, given computational limitations and other constraints.

Now, according to the pluralistic conception of human rationality presented in section 4.3, in different spheres of human life, different standards of rationality should apply. In most real life situations, due to incomplete information and other practical constraints, non-monotonic patterns of reasoning, which rely extensively on prior belief and external information, will for the most part deliver optimal results (this is a mix of the Panglossian and the Apologetic positions). But, in certain contexts, in particular contexts of scientific inquiry, it becomes important to counterbalance the human tendency towards doxastic conservativeness to facilitate the discovery of novel facts. Scientific methodology can be viewed as providing much of the cognitive scaffolding required to allow human agents to deviate from their more spontaneous cognitive tendencies (those that 'make sense' outside scientific investigation) in scientific contexts (so this is essentially a Meliorist position).

One of the claims defended in Chapter 4 was that the deductive method forces an agent to consider all possibilities, not only her 'preferred models', which is already a form of debiasing in the appropriate contexts. The claim to be defended in this chapter is that formal languages and formalisms are particularly useful tools for blocking the interference of external information and prior belief, in such a way that doxastic conservativeness can be countered in situations where this would be advantageous – most notably some (but not necessarily all) situations of scientific inquiry. A well-designed deductive formalism automatically forces the agent to consider all relevant possibilities, not only those compatible with her 'preferred models'. Thus, formal languages and formalisms are presented as possible *debiasing devices* in these specific domains of inquiry. The presupposition is that, at least in the case of scientific inquiry, excessive doxastic conservativeness can indeed be a hindering bias.

I start with very general philosophical considerations on what it means to formalize (sections 7.1.1. and 7.1.3), which will allow me to formulate the debiasing thesis more precisely. I then elaborate on how de-semantification and sensorimotor manipulation of the notation are combined for the debiasing effect to come about (section 7.1.2). In section 7.2, I adopt again an empirically informed, cognitive perspective to discuss the work of Houdé and collaborators on inhibition training, and to draw some implications from my debiasing account of formal languages and formalisms for the currently popular dual-process model of human cognition.

7.1 FORMALIZING TARGET PHENOMENA

7.1.1 The formalization process

Thus, in order to discuss the ways in which a formalization can have a debiasing effect, let me first offer a very general, schematic picture of what it means to formalize. As argued in section 3.3.1, I do not endorse the widespread view that formalization is a *translation* from portions of ordinary language into some formal language or formalism. Instead, I defend the view that formalization engages directly with the target phenomenon in question, without the detour via ordinary languages. But what *are* the target phenomena of formalizations? In a sense, getting clear on this question is both the most difficult and the most crucial step for a philosophical analysis of what it means to formalize. 'Target phenomenon' is merely a convenient, entirely general blanket term, used simply to refer to whatever it is that a formalization is a formalization *of*.

Now, when it comes to physical theories – for example, Maxwell's theory of electromagnetism discussed in section 6.2.1 – it would seem that the target phenomenon is a straightforward portion of reality, namely electromagnetic phenomena (but, even there, the ontology of physical reality is not nearly as unproblematic as one might think). But, soon enough, such (apparently) straightforward answers are no longer available, even in (seemingly) simple cases. For example, what is the target phenomenon of Peano arithmetic (PA)? Is it the *series* of the natural numbers tout court, or is it the *structure* consisting of this series *and* the distinguished operations of successor, addition, and multiplication? Different answers seem possible here, and yet it is clear that these are two quite dissimilar target phenomena. As we consider yet different kinds of formalizations (e.g., the formalization of historical logical theories, the formalization of correct canons of reasoning and arguing within a logical system, the formalization of rational agency with decision theory, etc.), it becomes evident that the exact (ontological) status of several target phenomena is a tricky issue.

Even if a satisfactory answer can be given to the 'what?' question, we are still left with the thorny issue of *epistemic access* to the target phenomenon of a formalization. Naturally, in order to formalize a given phenomenon, at least some degree of epistemic access must be presupposed on the part of the formalizer. However, as famously argued by Benacerraf (1973), while we should have epistemic access to the objects that our theory (and, in particular, formalization) refers to, in many significant cases (crucially for

Benacerraf, *mathematical* knowledge), a convincing story regarding our epistemic access to such entities is not forthcoming.

Arguably, the issues of epistemic access and the ontological status of target phenomena are most fruitfully discussed on a case-by-case basis; here, though, the goal is to articulate a general, schematic account of what it means to formalize, so the reader should not expect definitive answers at this point. For now, it is sufficient to flag these two issues as real problems for any account of what it means to formalize: the quasi-metaphysical issue of what the target phenomenon in fact is, and the epistemic issue of accounting for our access to it (even if only partial).

Now, having criticized the view that formalization is a process of translation from portions of ordinary language into a formalism (Chapter 3), I actually do endorse the claim that a formalization is, nevertheless, essentially a form of 'translation', but, more specifically what can be described as 'conceptual translation'. In effect, I submit that, at least in most cases, it is best to view the target phenomenon of a formalization as a thoroughly *intensional, conceptual entity*: theories, concepts, ideas. On this approach, the target phenomenon of, e.g., Peano arithmetic is not the free-standing, independently existing series of the natural numbers but rather our *theory* about this entity.[1] This does not mean that the 'Platonic' existence of such a series must be denied, nor does it mean that our theory of arithmetic is a purely social construction/ convention; it simply means that, *qua* target phenomenon, the starting point for a formalization/axiomatization of arithmetic is our *theory* of arithmetic itself. Naturally, this theory requires some medium to be expressed and communicated, such as a mixture of ordinary language and mathematical notation; but, again, while these expressions are required for epistemic access, the immediate object of formalization is not these portions of (ordinary) language. Thus, complementing the picture in 3.3.1, rather than

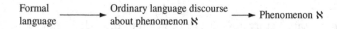

Formal language   ⟶   Ordinary language discourse about phenomenon ℵ   ⟶   Phenomenon ℵ

---

[1] Naturally, the Socratic exercise could be continued with the question, 'But what is a *theory?*' In the present context we will rely on an informal notion of theory, as a proposed account purporting to *explain* a given phenomenon.

what we in fact have is:

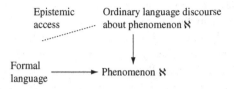

(The arrows indicate the relation of A having B as its target – A 'being about' B, so to speak.) In effect, the claim that target phenomena of formalizations are predominantly theoretical, conceptual, entities alleviates the pressure of the Benacerrafian challenge; epistemic access to the objects of formalizations comes almost 'for free' if the latter are viewed as belonging to the realm of the theories we actually entertain. The main challenge now seems to be one of objectivity, but it is not a challenge proper to formalization; any account of the objectivity of (scientific) theories in general would in principle be addressing this very challenge. Similarly, the challenge of accounting for the intentional content of an informal theory with respect to the objects it is about (in some suitable sense of 'about') still remains, but again it is deferred to a more general discussion beyond the specific topic of formalization.

Thus, in what follows I will take it as established that, generally speaking, the target phenomenon of a formalization is an *intensional* entity: a theory, a theoretical explanation of a given phenomenon, a concept. This also means that the connection between formalization and the object(s) it is about is typically mediated by the informal but theoretically robust account of the object(s) in question which is the target phenomenon; thus, there is no such thing as a theoretically neutral formalization. And, just as a theory can be descriptive or prescriptive, the same holds of a formalization.

This claim can be illustrated by means of some examples. Take Maxwell's theory of electromagnetism (section 6.2.1): in the first instance, his target phenomenon was Faraday's concept of 'lines of force', not electromagnetic phenomena as such. Similarly, in Adams's formulation of a semantics for conditionals in probabilistic terms (section 6.2.2), what was being formalized was Ramsey's concept of degrees of belief. The axiomatization of arithmetic, in turn, arguably takes as its target phenomenon our *theory* of arithmetic, not the series of natural numbers as such. We could multiply the examples here, but the main idea is that the starting point for a formalization is typically a theoretically robust though informally formulated concept or theory.

But how does one go about a formalization?[2] A formalization will typically not aim at including each and every aspect of its target phenomenon (the informal theory it takes as its starting point); choices will be made as to what aspects of the theory will receive a counterpart in the formalization, and what aspects will be left out (just as with any modelling enterprise). Moreover, the formalization may contain elements which do not correspond to anything in the target phenomenon, and instead are meant as convenient devices within the formalism (potentially making it simpler, easier to work with, etc.). Shapiro (1998) introduced the terms 'representor' and 'artefact' to refer respectively to elements of a formalization which do and do not have counterparts in its target phenomenon; we may also want to add the term 'left out' for the elements of the target phenomenon which are not represented in the formalization.

The 'representation' terminology clearly hints at what was described in section 3.2.1 as the *expressive* function of formal languages, which was nevertheless presented as not being the main factor behind the actual cognitive impact of using formal languages when reasoning. Is there an inconsistency here? Not necessarily, given that there are different stages in a process of formalization, which can be schematically represented as:

$$\text{Phenomenon } \aleph \xrightarrow{\text{F-in}} \boxed{\text{Formalization}} \xrightarrow{\text{F-out}} \text{Phenomenon } \aleph$$

The F-in stage corresponds to the expression of relevant features of the target phenomenon in the language of the formalization (typically, a logical or mathematical formalism), and at this stage some sort of structural similarity is desirable so that the fit between phenomenon and formalization is good enough. At this stage, concerns of expressivity are relevant, as they may play a *justificatory* role for the adequacy of the formalization.[3] Now, once the formalism has been designed and there is sufficient epistemic confidence in its capacity to characterize the target phenomenon, what occurs 'inside' the formalization should ideally make no appeal to external intuitions related to

---

[2] One of the most illuminating accounts of a process of formalization is Dedekind's famous 1890 letter to Keferstein, where he narrates 'the train of thought, which constitutes the genesis of [his] essay' (Dedekind 1890: 99). The essay in question was his reply to Keferstein's criticism of Dedekind's seminal work on the notion of number, *Was sind und was sollen die Zahlen?*

[3] However, a well-designed formalization may consist precisely in a creative, unexpected embedding of the target phenomenon onto a formalism which is not obviously structurally similar.

the target phenomenon. Indeed, at this stage, the formalism should be treated purely as a calculus, as this stance enhances its power to make non-trivial predictions – a true logic of discovery. But once conclusions have been drawn inside the formalism, they must be transferred back to the target phenomenon (the F-out stage), that is if the goal is to gain knowledge about the phenomenon (rather than investigating properties of the formalism as such – see section 7.1.3); here again, some degree of expressive fit between formalism and target phenomenon is to be desired.

Thus, at the F-in and F-out stages, some degree of expressive adequacy is to be expected of the formalization, in particular regarding the crucial issue of epistemic justification for the formalism as an adequate characterization of the target phenomenon in question.[4] Moreover, it is worth emphasizing that F-in and F-out are essentially processes of *conceptual* analysis: it could not be any different, as they concern the points of contact between the informal, conceptual realm of the target phenomenon and the formal, technical realm of the formalization.[5]

Nevertheless, when reasoning *inside* the formalism, expressive aspects must be disregarded as much as possible, so as to counter the 'fundamental computational bias' of bringing in prior belief when reasoning; at this point, emphasis should be laid on the *calculative* aspect of a formalism. In other words, while superficially it may seem that expressivity and calculability are in tension with one another as desiderata for formalization, in fact they *complement* each other at *different stages* of the process. However, the claim defended here is that, in terms of the cognitive boost afforded by a formalization, the calculative aspect is more significant in its mind-altering dimension, allowing us to run a 'different software', as it were (one that is not partial towards prior beliefs). So, let us now look more closely into the debiasing effect of reasoning with an adequate formalization.

7.1.2 Surprised by logic

The eighteenth-century mathematician and philosopher Jean d'Alembert is reputed to have said: 'Algebra is generous; she often gives more than is

---

[4] Given the schematic nature of the present analysis, I will not elaborate further on what it means for a formalization to be adequate, as it would seem that case-by-case analyses are to be preferred. But I would like to flag this issue as significant and yet under-appreciated. When is a formalization an adequate characterization of its target phenomenon? Brun 2003 and Baumgartner and Lampert 2008 provide some of the rare discussions of this important topic.

[5] P. Smith 2010 and Andrade-Lotero and Dutilh Novaes 2012 discuss these points of contact in terms of Kreisel's squeezing argument; we will come back to this topic in section 7.1.3.

asked of her.' Staal (2007b) has noted that the observation also holds of formal languages; indeed, formal languages have the generosity of often supplying us with unexpected, surprising results. How come?

Now, when reasoning *inside* the formalism, the less external beliefs and intuitions regarding the target phenomenon are appealed to, the higher the chance of arriving at a non-trivial result (prediction). Both Frege and Hilbert had emphasized the importance of blocking presuppositions and external intuitions for logical and epistemological reasons, but, as we have seen in the survey on human reasoning in Chapter 4, there are also *cognitive* reasons why sneaky presuppositions tend to crop in. Our strong tendency towards seeking confirmation for the beliefs we already hold – our doxastic conservativeness – stands in the way of the discovery of truly novel (unknown, unexpected) facts about a given phenomenon.

How are de-semantification and sensorimotor manipulation of a formalism combined for the debiasing effect to occur? Here is a somewhat oversimplified but hopefully illuminating account of this process: de-semantification 'turns off' the fundamental bias of bringing in prior belief while reasoning, insofar as it blocks or at least attenuates semantic activation; then, a new modus operandi can establish itself, one which is not partial towards prior beliefs, namely the sensorimotor (mostly visual) manipulation of the notation (which, it bears emphasizing, must be *trained for* to be mastered). Because the latter resorts to cognitive systems and processes that are quite different from those involved in our spontaneous tendency to rely on prior belief, the likelihood of the fundamental bias cropping in again is probably much lower.

To be sure, at least half of this story – the de-semantification half – is compatible with an internalist account of manipulating notations, so those who do not endorse the externalist account need not reject the picture presented here altogether. Still, it seems to me that the case for the debiasing effect of formalisms becomes even stronger if externalization of reasoning is part of it; sensorimotor processing is simply not likely to be tendentious towards prior beliefs. Similarly, it is possible to reason 'formally' with a given formalism without the de-semantification step, as in Frege's ideography, yet de-semantification seems greatly to facilitate the process of blocking substantive intuitions concerning the target phenomenon. Thus, in a slogan: de-semantification 'turns off' the fundamental computational bias, and sensorimotor manipulation of the notation 'turns on' a different software.

Naturally, this way of describing the debiasing effect of formalization is essentially metaphorical, not intended to characterize with absolute accuracy the actual cognitive processes of an agent reasoning with a formal language or formalism (i.e., Marr's 1982 'implementation level'). But it arguably captures the gist of what seems to be the considerable potential for a debiasing effect when operating with formal languages and formalisms.

The claim that the calculative aspect of reasoning with a formalism is a potential source of surprises is of course in stark contrast with Wittgenstein's famous Tractarian thesis: 'Hence, there can never be surprises in logic' (6.1251). What is more, Wittgenstein ascribes the lack of surprises in logic precisely to what is described here as one of the main factors for the emergence of surprising results with (logical or otherwise) formalisms:

6.1262 Proof in logic is merely a mechanical expedient to facilitate the recognition of tautologies in complicated cases.

I have suggested (Dutilh Novaes 2010a; 2010b) that what is missing in the Tractarian picture is the *subject* who actually draws the inferences. The claim that there can be no surprises in logic stems from an undue disregard for the epistemic act of actually unpacking the information contained in the premises when drawing a (deductive) inference. Before the act of unpacking, the information contained in the premises is merely 'virtual information' which must be actualized in order to become truly available, and this is exactly what the act of drawing an inference is able to accomplish. And, *contra* Wittgenstein, the act of unpacking information by means of proofs and deductive inferences may well deliver surprises.

In fact, the claim here is precisely that one of the main goals of a formalization is to arrive at a situation of *mismatch* between some of one's original presuppositions and beliefs concerning the target phenomenon and the results and predictions of the formalism – to be surprised by logic. There is, however, a catch here: such mismatches may mean either that the formalization is indeed informative in that it exposes features of the target phenomenon which are not apparent to the 'naked eye' (pursuing Frege's analogy of a formalism as a microscope), i.e., on the basis of conceptual analysis alone; or that the formalization is *defective*, i.e., that it has not captured the target phenomenon adequately. So the same outcome may be a sign of adequacy as well as of inadequacy. Analogously, a formalization which confirms all presuppositions and hunches one already had concerning the target phenomenon is on the one hand 'adequate', in that the fit with the phenomenon seems optimal, but on the other hand it is non-informative in that it did not reveal anything novel about the latter. We could refer to this as the *paradox of adequate formalization*.[6]

Here again, there does not seem to be a general solution to the paradox: case-by-case analyses would be required to determine whether a situation of mismatch is a sign of a 'good' or of a 'bad' formalization. But a general methodological recommendation seems to be in order; if one has strong independent reasons to attribute a high level of epistemic reliability to the formalization, in particular at the F-in and the F-out stages, then the recommendation is to be open-minded about the novel prediction being made by the formalism, i.e., to prevent doxastic conservativeness from settling in immediately. Perhaps what appears to be a truly counterintuitive prediction at first in fact ties in nicely with other 'loose ends' of the theory, and then goes on to receive further corroboration once the idea is put in a different light (such as the idea that no physical medium such as a hypothetical ether was required for the propagation of light, discussed in section 6.2.1).

It may be thought that I am making the naive suggestion that operating with a formalism is a purely mechanical endeavour in that every step is fully determined by the rules of transformation, such as in a Turing machine or an arithmetic algorithm. If this were the case, proving theorems within the formalism would be an utterly trivial endeavour, which is clearly absurd. A significant amount of 'insight and ingenuity' is usually required in the formulation of non-trivial proofs within a formal system (as suggested in section 3.3, in a demonstration, the proof itself is the unknown value). The main reason why this is the case is that typically, from any given statement within a formal system, a range of different transformational steps are allowed, which of course leads to a combinatorial explosion of possibilities. Thus, there is still an important amount of creativity and insight involved in producing a non-trivial proof, often pertaining to, for example, clever renaming of schematic letters, unexpected instantiations of axioms, etc. Indeed, even automated theorem-provers must contain some heuristic features in the program, as 'brute force' alone (i.e., mere combinatorial procedures) leads to an explosion of trivial, uninformative theorems.[7] So, perhaps contrary to Leibniz's original vision, reasoning with a formalism is in fact not simply a matter of combinatorial analysis.

---

[6] I have discussed these ideas in more detail (Dutilh Novaes 2011c), specifically with respect to the formalization of historical logical theories, but the observations there hold of formalizations more generally.

[7] Pelletier has some very interesting papers on the philosophy of automated theorem proving (e.g., Pelletier 1991; 1998).

But the formalism does block certain moves which might have seemed intuitive from the point of view of the target phenomenon, and this is sufficient for the debiasing effect to occur. In fact, this feature is not proper to operating with formalisms; it is one of the cornerstones of the method of mathematical proofs. Here is an anecdote to illustrate the point: the renowned logician/mathematician Jeff Paris once spent weeks trying to prove a statement which he was sure was true; multiple computer simulations also seemed to converge towards the truth of the statement. But the proof was simply not forthcoming. Then, in desperation, he set out to look for counterexamples to the statement, and found one within less than one hour. He was strongly biased towards the truth of the statement in question, and yet the method of mathematical proof simply did not allow him to formulate a confirmation for his (false) belief.[8]

7.1.3 Applying logic

Over the years, I have been presenting the main ideas of this book to different audiences, and one objection that is often put forward is that my account of formal languages as debiasing devices simply does not correspond to the actual current practices of most logicians. It is frequently pointed out to me that logicians do not primarily reason *within* logical systems so as to investigate external target phenomena, and that instead they investigate properties *of* logical systems, in particular, soundness, completeness, decidability, etc. This is of course quite true, and is essentially a result of the meta-turn taken in logic in the 1920s and 1930s (as discussed in Chapter 3).

In light of these observations, it may be useful to offer some considerations on the scope of the claims defended here, in particular regarding the distinction between using logic to study extra-logical phenomena and studying logical systems as such. A useful terminology to address this distinction is introduced by F. Paoli, specifically in connection with the issue of logical constants, as an adaptation of the originally Quinean distinction between *transcendent* and *immanent*:

According to Quine, in fact, logical connectives are immanent, not transcendent. There is no pretheoretical fact of the matter a theory of logical constants must account for; rather, the vicissitudes of a connective are wholly internal to a specified formal language, to a given calculus. There is nothing, in sum, that precedes or transcends formalization, no external data to 'get right'. (Paoli 2003: 542)

---

[8] I thank Jeff Paris for sharing this story in a personal communication.

We can generalize the main idea beyond the specific case of logical constants in the following way: on the immanent approach, a logical system is seen as an independent, free-standing (mathematical) entity, with no particular ties to any phenomena external to itself. The task of the logician is to investigate the properties of logical systems as independent entities. On the transcendent approach, by contrast, a logical system *is* about some external phenomena, i.e., it does seek to capture and characterize something external to itself.

Priest offers similar considerations in the context of his discussion of logical pluralism; he speaks of *pure and applied logics*, in an analogy with pure and applied geometries.

there are many pure logics . . . Each is a well-defined mathematical structure with a proof-theory, model theory, etc. There is no question of rivalry between them at this level. This can occur only when one requires a logic for application to some end. Then the question of which logic is right arises. (Priest 2006: 195)

As should be evident by now, when speaking of the debiasing effect of logical formalisms, it is the realm of *applied logics* that is in question, i.e., the realm of logical systems being used to characterize and investigate a given extra-logical phenomenon. (Similar considerations would hold of the debiasing effect of mathematical formalisms in applied mathematics and elsewhere.) Similarly, we are interested in the *transcendent* connection between a logical system and an external phenomenon that it seeks to characterize. This is not to deny the importance of the pure/immanent perspective in logic, but, as discussed in Chapter 3, the predominance of metalogical investigations in the practices of logicians since the 1920s has had the inconvenient consequence of deflating the significance of logical systems as tools to reason *with*, about something *else*.

But this has not happened across the board. Take Carnap's *The Logical Syntax of Language*, for example: while Carnap may seem to adopt an 'immanent' stance there ('In logic, there are no morals'), in practice the main motivation for his investigations is arguably the idea of logical systems serving as tools for scientific and intellectual inquiry (in philosophy in particular). A significant portion of analytic philosophers then went on to pursue the Carnapian ideal and made extensive use of logical tools to investigate philosophical matters (Kripke and Lewis are two names that immediately spring to mind). But *logicians* themselves became increasingly uninterested in the possible connections between their logical systems and external phenomena, in such a way that 'applied logic' became mostly practised by people who did not self-identify as logicians.

Thus, the upshot is that the potential debiasing effect of operating with formal languages and formalisms is arguably not particularly felt in the current work of those who self-identify as 'logicians' (as these mostly adopt a 'pure logic' perspective),[9] but rather in areas where formalisms are applied to investigate specific phenomena: economics, different branches of philosophy (formal epistemology, metaphysics), physics, biology, etc. Naturally, it is primarily when the goal is to study a given external target phenomenon that it is appropriate to counter doxastic conservativeness about the phenomenon in question. True enough, it cannot be ruled out that, even for metalogical investigations, manipulating formalisms may have a debiasing effect; logicians may have hunches concerning the very properties of their logical systems, which then may or may not be confirmed by further formal analysis. But the present investigation pertains predominantly to uses of logical tools to investigate extra-logical phenomena: reasoning *with* the logical system rather than reasoning *about* it.

Adopting now the transcendent stance, the most pressing question for an account of how formal, technical frameworks can be applied to investigate informal, non-technical concepts and theories then becomes: how do the two realms 'latch on' to each other? What does it mean for a formal framework to *characterize* or *capture* a given informal theory or concept? There is a sense in which the fit will always be imperfect, i.e., there will always be 'gaps' (as suggested by Shapiro, see section 3.3.1), simply because these are two highly heterogeneous theoretical realms. But scepticism concerning the epistemic power of formal modelling is put under pressure by the sheer amount of successful cases of genuine novel knowledge about a given phenomenon being produced by means of formal modelling. At any rate, it seems undeniable that some sort of 'latching on' does take place, even if it is not entirely obvious how to formulate a precise account of the underlying mechanisms.

An illuminating account of how informal and formal frameworks can be coupled is offered by Peter Smith (2010) in terms of Kreisel's famous squeezing argument.[10] Consider a theoretically robust but informally formulated concept I ('I' for 'informal', not for 'intuitive'), typically with somewhat vague borders.[11] Assume that there is a technical, formal, and

---

[9] There are exceptions, of course, but even the so-called field of applied logics concerns for the most part the interface between logic and computer science.

[10] I here follow the presentation in Andrade-Lotero and Dutilh Novaes 2012, inspired by Smith's account.

[11] The emphasis on 'theoretically robust' is related to the arguments offered in section 7.1.1 for the claim that the immediate target phenomenon of a formalization is always a theoretical, conceptual entity.

precisely defined concept S, which can be predicated of (roughly) the same category of entities *e* as I, and which, as it turns out, seems to offer *sufficient* conditions for an entity to fall under I. Thus:

(1) If *e* is S, then *e* is I.

Assume moreover that there is another technical, formal, and precisely defined concept N which uncontroversially provides *necessary* conditions for something to fall under I.
Thus:

(2) If *e* is I, then *e* is N.

Given these two implications, 'the extension of I – vaguely gestured at and indeterminately bounded though that might be – is at least sandwiched between the determinately bounded extensions of S and N' (Smith 2010: 23). Now, what makes squeezing arguments truly interesting is that, in some cases at least, there is a third implication whose truth can be established by means of a *proof* (given that S and N are technical, mathematically defined concepts):

(3) If *e* is N, then *e* is S.

Given this third implication, the informal concept I is 'squeezed' between the two technical concepts, i.e., its extension is shown to coincide with that of the other two concepts. We thereby obtain a precise, technical formulation of the boundaries of informal, vague concept I.

This schematic formulation of the squeezing argument can receive different instantiations; in Kreisel's original formulation, I is an informal concept of validity, and what is thought to provide sufficient conditions for something to count as I is a *deductive system* for a (first-order) logic, while what would provide necessary conditions for something to count as I is the (standard) *semantics* for the logic. In such cases, (3) is simply a proof of completeness for the appropriate pair of deductive system-semantics.

But the heart of the matter is that the cogency of premises (1) and (2) in particular cases cannot be established purely by means of proofs (contrary to (3)), precisely because one of their elements is an informal concept, I; thus, their adequacy must be argued for essentially on *conceptual* grounds.[12] And

---

[12] See Andrade-Lotero and Dutilh Novaes 2012 for a discussion of the potential inadequacy of (2) in specific cases.

yet the adequacy of a formalization of I by means of S and N relies crucially on the cogency of the relevant instantiations of (1) and (2) in specific cases. The point being made here is not that the fit will never be sufficiently adequate, but rather that adequacy cannot simply be taken for granted; it must be argued for. There must be good reasons to maintain that specific instantiations of (1) and (2) hold in particular cases.

Admittedly, this schematic version of the squeezing argument offers one among different possible models of how formalisms connect with their target phenomena. It most likely does not exhaust occurrences of these connections in specific cases, but it offers a concrete framework to discuss the matter. Similar analyses could presumably also be carried out with other models, so as to illustrate different aspects of the relations between formalisms and target phenomena.

At any rate, the main conclusion of this section is that there are reasons to believe that it *is* possible for informal and formal theoretical realms to 'latch on' to each other, possibly with different degrees of closeness of fit. But these connections must be properly motivated, and this is to be done essentially on conceptual, informal grounds – a conceptual translation.

7.1.4 Conclusion

In this section, I approached the concept of formalization from a very general, schematic perspective. I suggested that the target phenomena of formalizations are typically intensional, theoretical entities – theories, concepts. I then presented an account of how de-semantification and sensorimotor manipulation of the notation are combined to bring about the debiasing effect of formal languages and formalisms, so that surprising conclusions about the target phenomenon of the formalization are more likely to emerge. Furthermore, I specified the scope of my claim concerning the debiasing effect of formalisms: the domain of *applied* logic, i.e., logical formalisms being used to investigate external, extra-logical target phenomena. I also offered considerations on how formalism and target phenomenon can 'latch on' to each other, on the basis of a schematic version of Kreisel's squeezing argument. My conclusion is that it *is* possible for a formal framework to characterize and capture an informal theory/concept (with various degrees of fit), but that there will always be a 'gap' between phenomenon and formal model. Moreover, the adequacy of a formalism cannot be taken for granted; it must be argued for on conceptual grounds.

7.2 DEBIASING AND COGNITION

In this section, I discuss the (to my knowledge) only research programme focusing specifically on debiasing strategies in the domain of reasoning, namely the work of Houdé and collaborators on inhibition training. I then turn to dual-process theories of cognition and argue that they (in their classical versions, in any case) can simply not accommodate the account of the debiasing effect of operating with formal languages and formalisms proposed here.

7.2.1 Inhibition training

While reasoning biases have been extensively studied by psychologists, possible *de*biasing strategies specifically for reasoning have received proportionally little attention.[13] Evans, et al. 1994 provides a notable but fairly isolated exception within the psychology of reasoning, identifying a significant (though not unlimited) debiasing effect in instruction on the concept of logical necessity. A more systematic investigation of debiasing mechanisms can be found in the work of Houdé and Moutier and their collaborators. They conducted several studies to investigate the effect of *inhibition training* in a range of different reasoning tasks. Here I discuss in particular their results on matching bias (previously discussed in section 6.1.3) in Wason selection tasks (previously discussed in section 4.1).

One of the most interesting aspects of the work of Houdé, et al. is their emphasis on the effects of *training*. As discussed in Chapter 4, the psychology of reasoning tradition has for the most part neglected the influence of instruction and formal education on reasoning.[14] Elqayam and Evans offer very appropriate observations on this omission:

The standard practice in the psychology of reasoning, at least until the past decade or so, was as follows. You draw a sample of participants, specifically excluding any who have had formal training in logic. You present them with problems which are either abstract, or designed in such a way that any influence of real world beliefs can only be interpreted as a bias, as it is orthogonal to the logical structure ... The argument encouraged by Inhelder and Piaget and numerous philosophers is essentially this: (classical) logic provides the laws of rational thought in all

---

[13] By contrast, debiasing has been extensively studied in the field of judgment and decision-making, as mentioned at the beginning of this chapter.

[14] One exception is arguably the work of 'Meliorist' K. Stanovich, who has also worked extensively on education (in particular with respect to reading).

çontexts.[15] People are rational. Therefore, logic must be built into people's heads in some innate and *a priori* manner ... Why on earth, for example, should our notion of rationality exclude *learning*? Why not start with the observation that people can become expert reasoners in law, medicine, science, engineering and so forth, noting that in every case they spend many years in specialized training to achieve this level of expertise? Why not focus on expert reasoning and how it is acquired? But no, we have spent the past half a century instead studying naïve participants with novel problems, resulting in a Kuhnian crisis as the field struggled to throw off the shackles of logicism. (Elqayam and Evans 2011: 243)

Indeed, learning and training have remained almost completely out of the picture, but, as argued in section 4.3, there are many reasons to think that strict logical, deductive reasoning primarily emerges upon specific training. The fact that these studies are invariably conducted with participants having a homogeneous educational background (typically, undergraduates in universities in Western Europe and North America) does not allow for the separation of possible influences of educational factors, and the few studies conducted with participants with a significantly dissimilar educational background (Luria 1976; Scribner 1997; Counihan 2008b) suggest that their reasoning strategies tend to be radically different.

Thus, the focus of Houdé and collaborators on the effects of a specific kind of training offers a refreshing change of paradigm. Rather than merely synchronic, their studies typically include at least three stages: pre-training tests, short training (of different kinds for different groups), and post-training tests. Nonetheless, they seem to have inherited the bulk of the Piagetian presuppositions criticized by Elqayam and Evans in the passages above. It would seem that, on their picture, logic really is 'built into people's heads in some innate and a priori manner', but human reasoners tend to deviate from the traditional deductive canons due to the *interference of conflicting cognitive processes*. Inhibition consists precisely in the *suppression* of these conflicting processes, so that logical competence can overrule them and prevail.

Since Piaget, it has been known that cognitive development culminates in formal logical thinking,[16] generally assumed to be crucial for human adaptivity. (Moutier, et al. 2006: 166)

These findings seem to show that flawed performance on deductive logical tasks does not necessarily signal any deficit in deductive logical ability per se, but can be

---

[15] In a commentary on Elqayam and Evans 2011, I have argued (Dutilh Novaes 2011d) that this is essentially a Kantian legacy.

[16] This is controversial at best, and possibly plainly false, as argued in section 4.1, above.

caused by uninhibited biases (e.g., matching bias and belief bias). (Moutier, et al. 2006: 167)

So, several concurrent reasoning strategies might compete at any time, even during adulthood, in such a way that perceptual responses often override logical ones, and cognitive inhibition turns out to be the key that opens the door to deductive logic. (Houdé and Tzourio-Mazoyer 2003: 509)

Let me now discuss their studies on matching bias. Recall that matching bias refers to the tendency participants have to select items that are mentioned/signified by the very terms used in the task material (typically, a conditional rule in some formulation of the Wason selection task). As presented in section 6.1.3, matching bias clearly is essentially a *lexical*, semantic phenomenon, arguably related to the automaticity of semantic activation; Evans (1998) also characterizes it as a lexical phenomenon. Houdé and collaborators, as well as other researchers such as Prado and Noveck (2007), however, conceptualize matching bias as a *perceptual* phenomenon (hence the reference to 'perceptual responses' in the passage just quoted).

It is quite puzzling why they should view matching bias as a perceptual phenomenon, and the only explanation I can offer is that the task materials they work with are often conditionals mentioning shapes and colours, such as: 'If there is not a red square on the left, then there is a yellow circle on the right' (Houdé, et al. 2000: 721). But in the original formulation of the Wason selection task, for instance, where mention is made of (odd or even) numbers and of (vowel or consonant) letters, it is hard to see how selecting the cards signified by the terms used in the conditional rule (e.g., [A] and [3] if 'vowel' and 'odd number' are mentioned) could be adequately characterized as a *perceptual* phenomenon.

One of the presuppositions at play here seems to be the good old rationalist opposition between perception (a lower cognitive faculty) and reasoning (a higher cognitive faculty), and debiasing would of course pertain to the realm of higher cognitive faculties.[17] However, if the account of reasoning with notations presented in Chapter 5 is correct, it is *precisely* the engagement of sensorimotor processes that facilitates the debiasing effect of operating with formal languages and formalisms (a point further discussed in the next section). At any rate, the characterization of reasoning

---

[17] 'This is a good example of abstraction – a controlled and analytical kind of processing that occurs when our brain can break away from perception (which relies more heavily on automatic and heuristic processes)' (Houdé and Tzourio-Mazoyer 2003, 508–9).

biases (or even of matching bias alone) specifically as a perceptual phenomenon seems highly problematic.

But Houdé and collaborators do reject another classical dichotomy, namely that between emotion and reason. They incorporate Damasio's (1994) suggestion that emotion plays an important role in sound reasoning, and describe the kind of training which elicits the highest rate of 'correct' responses in the post-test as 'logicoemotional'.[18] While there is much to be commended in rejecting old dogmas, it seems that describing the inhibition effect observed in their experiments as having an emotional component is a mistake, as I will argue now.

First, here is a short description of the basic experimental set-up:

To study the perceptual matching bias, we hypothesized that adults, like children during the acquisition of elementary cognitive abilities, have two competing strategies in their mental workspace – one logical and one perceptual – but they have trouble inhibiting the perceptual one. This is not really a question of mental logic *per se*, but one of executive control, in this case cognitive inhibition. To show this, we used three training conditions: inhibition of the perceptual strategy (including emotional warnings about the error risk), logical explanation and simple task repetition (which corresponds to practice). Only inhibition training proved effective in reducing errors: the success rate increased from 10 per cent to above 90 per cent. This means that an executive inhibitory mechanism is what these adults were lacking, not logic or practice. (Houdé and Tzourio-Mazoyer 2003: 509)

And here is an excerpt from the 'logicoemotional' training offered to participants in a classical Wason selection task study; the italics were added by the authors themselves to emphasize what they took to be the 'emotional warnings' of the training:

In this problem, the source of the error lies in a habit we all have of concentrating on cards with the letter or number mentioned in the rule . . . and not paying attention to the other cards . . . Thus, the goal here is (1) *not to fall into the trap of the two cards A and 3 mentioned in the rule* and (2) to consider all of the cards, A, D, 3, 7, one by one, by imagining the number or the letter it might have on the back to see whether these cards can make the rule false . . . To help you understand, let's consider the different answers and *eliminate the wrong ones – the ones that make you fall into the trap* – to find the right answer. (Houdé, et al. 2001: 1488)

The so-called logicoemotional training also included a hatched area on top of the box used to sort the right from the wrong answers. By contrast, the

---

[18] 'Does the human capacity for access to deductive logic depend on emotion and feeling? . . . The error-to-logical shift occurred in a group that underwent logicoemotional training but not in the other group, trained in logic only – a "cold" kind of training' (Houdé, et al. 2001: 1486).

so-called purely logical training, which in their studies systematically failed to produce any significant increase in the 'logically correct' responses, was as above, but excluded both the so-called 'emotional warnings' shown in italics as well as the hatched box (this last element will be disregarded in the present discussion).

The first thing to notice is that the so-called logicoemotional training is much more thorough and detailed, literally taking the participant 'by the hand' and showing her how to proceed, whereas the purely logical training is much more schematic; the former contains all the elements of the latter, but much more. This fact alone would already go a long way towards explaining at least partially why the more thorough training is significantly more effective.

However, what seems most questionable in this conceptualization of the experiment is the attribution of an 'emotional' character to the additional elements in the so-called logicoemotional training. I submit that, rather than emotional, these warnings are of a *meta-cognitive*, cognitive monitoring nature; they reveal to the participant what her spontaneous cognitive reactions are likely to be, why they constitute an obstacle for the successful completion of the reasoning task at hand, and what to do to circumvent them. Now, it is immediately obvious that such warnings are likely to be very effective debiasing strategies. In effect, that meta-cognition and meta-level awareness of some key concepts (such as that, in deductive settings, all options must be considered) are likely greatly to enhance performance in deductive settings is indeed one of the implications of the view defended here, namely that 'logical competence' comes first and foremost from training and education.

Thus, I strongly disagree with the authors' own conceptualization of the experiments and results, in particular their account of matching bias as perceptual and of meta-cognitive training as emotional, and their endorsement of the Piagetian idea that logical competence really is in people's heads, albeit buried under powerful biases (it simply needs to be properly 'activated'). But the results themselves are extremely interesting, in particular in that they highlight the powerful effect of (cognitive monitoring) training to elicit responses aligned with the traditional canons of deduction.[19] Recall

---

[19] Many of the Houdé, et al. studies also focus on neural aspects of these experiments, but, given my reservations towards their conceptualization of the set-up, I could not find anything sufficiently conclusive and convincing in their neuro-imaging analyses which could further inform the present discussion. However, a study by Goel and Dolan (2003) found a correlation between participants who are able to overcome belief bias and activation of brain areas usually associated with cognitive monitoring, which offers support to my interpretation of the Houdé, et al. results as concerning meta-cognition and cognitive monitoring.

that, in Chapter 4, it was suggested that the deduction game is not a game that humans spontaneously play, but we can certainly *learn* to play it with the appropriate training.

My main claim is thus that these studies on bias inhibition resort first and foremost to a *meta-cognitive level* in order to obtain the debiasing effect.[20] Participants are explicitly warned that their spontaneous cognitive patterns are likely to lead them astray in the specific tasks at hand, and told what to do to suppress these otherwise spontaneous processes. The authors take their results to lend support to the Piagetian idea that logic is 'in the head', but that it just needs the chance to emerge, in particular in that purely logical training does not seem to be sufficient to enable the participant to have 'access' to her 'logical competence'. However, the results are equally compatible, and perhaps even more so, with the claim that logical reasoning must be learned to be mastered, and precisely because it is so far removed from our spontaneous reasoning patterns (our 'biases'), recourse to debiasing mechanisms such as meta-cognitive warnings has a significant effect; succinct logical instruction alone is not sufficient. Hence, resorting to meta-cognition to elicit bias inhibition is one possible approach to debiasing, which at least in the limited setting studied by Houdé and collaborators seems to be quite effective.

The debiasing effect of formal languages and formalization as presented in Chapters 5 and 6 and in section 7.1.2 is of a very different nature; while in meta-cognitive training participants are instructed to monitor their cognitive processes carefully so as to be able to 'intervene', sensorimotor manipulation of notations following the rules of transformation of a system seems to work best precisely if the agent 'thinks' as little as possible about what she is doing (as well said by Whitehead).[21] But, naturally, nothing prevents the existence of *different* debiasing techniques, which may very well operate in radically dissimilar ways; there is no inconsistency in claiming that meta-cognition/cognitive monitoring as well as a combination of de-semantification and sensorimotor manipulation of the notation can both have a debiasing effect, but through very different cognitive channels. They may even be combined at different stages of learning processes: an emphasis on cognitive monitoring at the early stages, but gradually the processes in question become increasingly automatic.

---

[20] A different approach underpins their work on belief bias inhibition, exploiting the concept of negative priming (Moutier, et al. 2006).

[21] 'It is a profoundly erroneous truism ... that we should cultivate the habit of thinking what we are doing. The precise opposite is the case. Civilization advances by extending the number of important operations which we can perform *without thinking about them*' (Whitehead 1911: 61; emphasis added).

As suggested by the results of Evans, et al. (1994), straightforward instruction on the concept of logical necessity has a limited (immediate) debiasing effect; arguably, a combination of meta-cognitive training and emphasis on sensorimotor manipulation of the notation can be even more effective. However, these two approaches to debiasing go in opposite directions, and thus have entirely different implications for a currently very popular model of human cognition, namely the dual-process model; so let us turn to these implications in the next section.

7.2.2 Implications for dual-process theories of cognition

Thus, to recap, I have argued that the Houdé, et al. inhibition training resorts essentially to *meta-cognition* and *cognitive monitoring* to obtain a debiasing effect; debiasing would arise from increasing one's level of consciousness about one's own cognitive processes. By contrast, the debiasing effect of engaging sensorimotor processes when operating with formal languages comes from consciously thinking *less* about the cognitive operations in question, and from carrying out these operations 'mechanically', as it were. (It may be worth emphasizing again that operating with formalisms is a skill that must be trained for.) De-semantification is also a form of inhibition, namely the inhibition of semantic activation; but, on my account, inhibition of semantic activation does not come about through meta-cognition, and it should not require a special effort to be achieved. In short, the inhibition training presented by Houdé, et al. is related to higher cognition and executive control, while my account of debiasing through formal languages and formalisms goes in the opposite direction.

As may have become apparent in the previous section, the Houdé, et al. picture of conflicting, parallel cognitive processes is squarely in the spirit of *dual-process theories of cognition* (previously mentioned in section 4.3.3). In this section, these theories and their implications for the present analysis will be discussed in more detail. There is a vast literature on the topic, ranging from philosophy (of mind) to psychology, and within psychology, ranging from decision-making and social psychology to reasoning. But in order to focus on the essentials of the framework I will mostly be following the concise but comprehensive survey of Frankish (2010).[22] Its abstract already contains most of the elements needed for the present discussion:

---

[22] For a more general, state-of-the-art overview of the topic, see the collection of papers of Evans and Frankish (2009).

Dual-process theories hold that there are two distinct processing modes available for many cognitive tasks: one (type 1) that is fast, automatic and non-conscious, and another (type 2) that is slow, controlled and conscious. Typically, cognitive biases are attributed to type 1 processes, which are held to be heuristic or associative, and logical responses to type 2 processes, which are characterised as rule-based or analytical. Dual-system theories go further and assign these two types of process to two separate reasoning systems, System 1 and System 2 – a view sometimes described as 'the two minds hypothesis'. It is often claimed that System 2 is uniquely human and the source of our capacity for abstract and hypothetical thinking. (Frankish 2010: 914)

Dual-*system* theories can then be further specified by claims of precise *anatomic localizations* of each of the two systems.[23] Here, I focus on dual-*process* theories, but the points to be made naturally generalize to their stronger cousins.

There are several alternative characterizations of the type 1 v. type 2 dichotomy available in the literature; Frankish (2010: 922) presents a convenient table summarizing the main oppositions. For our purposes here, Table 7.1 contains the main features of each. Frankish's table does not contain 'logical' as one of the features of type 2 processes, but the capacity for 'logical reasoning' is typically viewed as a type 2 process. Indeed, recall that Houdé and collaborators endorsed the Piagetian idea of the availability of 'logical competence' in humans, but explained the 'poor performance' of participants in deductive tasks in terms of the interference of other dominating cognitive processes, which they described as 'perceptual' and unconscious – that is, a classical type 2 v. type 1 opposition. On their account, inhibition has a debiasing effect because it suppresses type 1 processes, thus making room for type 2 processes to emerge. The title of one of their papers (Houdé, et al. 2000) says enough: 'Shifting from the Perceptual Brain to the Logical Brain: The Neural Impact of Cognitive Inhibition Training'.

Although 'perceptual' is not among the items characterizing type 1 processes on Frankish's table, he readily recognizes this frequent association when commenting on how the two types map onto the everyday distinction between intuition and reason:

The core of dual-process theory is present in the everyday distinction between intuition and reason – the former immediate, *quasi-perceptual*, sensitive to subconscious cues and sometimes biased; and the latter slow, effortful, explicit and more cautious. (Frankish 2010: 915; emphasis added)

---

[23] In particular, see Goel 2007.

Table 7.1

Type 1	Type 2
automatic	controlled
non-conscious or pre-conscious	conscious
associative	rule-based
actual	hypothetical
concrete	abstract
non-verbal	language-involving
slow acquisition and change	fast acquisition and change

And, last but not least, on the classical picture (but which has been revisited in recent years),[24] 'cognitive biases are attributed to type 1 processes, which are held to be heuristic or associative, and logical responses to type 2 processes, which are characterised as rule-based or analytical' (Frankish 2010: 914).[25] In effect, the presumed conflict between biased and logical reasoning was one of the initial motivations for the development of some versions of dual-process theories, Evans's, in particular.[26] Thus, on the classical dual-process picture, debiasing is above all a matter of suppressing type 1 processes and thereby enhancing type 2 processes. And, given that type 2 processes are also described as characterized by a higher level of consciousness, the association between logical, rule-based reasoning and conscious, controlled processes establishes itself quite naturally. In a slogan: to reason logically, one has to think 'harder'.

Notice, however, that this picture is in stark contrast with the ideal of *mechanical reasoning*, which has been associated with the concept of logical reasoning at least since the seventeenth century. Recall (as discussed in Chapters 1 and 3) that to reason formally/logically is often viewed as closely related to *computing/calculating*, which in turn is viewed as reasoning with *no appeal to insight and ingenuity*: performing operations 'without thinking about them', in Whitehead's fitting terminology. Thus, we have two opposite conceptions of reasoning logically: one involving a high level of

---

[24] 'Likewise, many researchers now accept that it is wrong to characterise System 2 reasoning as uniformly abstract, rule-based and logical ... [M]any writers now accept that System 2 may fail to deliver normatively correct results. In the updated version of his heuristic–analytic theory, Evans assigns cognitive biases as much to analytic reasoning as to heuristic processes' (Frankish 2010: 921).

[25] This is clearly a component of the Houdé, et al. account (Houdé, et al. 2000).

[26] 'A second motivation was to account for an apparent conflict between logical processes and non-logical biases on deductive reasoning tasks, notably syllogistic reasoning, where participants have to evaluate the validity of an argument. Analysis of participants' responses suggested that two competing processes were at work: a logical process sensitive to deductive relations, and a "belief bias" that leads participants to endorse arguments with believable conclusions' (Frankish 2010: 916).

cognitive and meta-cognitive consciousness; and another involving precisely a *decrease* in levels of consciousness so as to enhance logical competence. It may well be that both processes facilitate logical reasoning, but clearly there is a tension here which must be further explored.

In fact, the account of the debiasing effect of operating with formal languages presented here simply does not fit neatly into the standard dual-process conceptions of human cognition. To be sure, there is a component of inhibition of automatic processes, namely the interference with semantic activation that de-semantification brings about, but this affects language-based reasoning, which in turn is often viewed as a type 2 process (see Table 7.1).[27] Moreover, this inhibition does not take place by means of executive control, as in the case of the inhibition training of Houdé, et al.; instead, features of the *external medium* (the notation) should prompt this effect.

More significantly, though, the emphasis on *externalization* of reasoning in order to enhance 'mechanical thinking', and on the engagement of *sensorimotor* processing makes operating with formal languages (on this conceptualization at least) look very much like a *type 1* kind of process (despite being rule-based, which is typically viewed as a characteristic of type 2 processes). In effect, appropriate training should turn the sensorimotor manipulation of notations into a largely (though not entirely – see comments on heuristics in section 7.1.3) *automatic* affair, not requiring conscious reflection at every step.

Now, on most dual-process approaches, the idea of type 1 processes being capable of bringing about a debiasing effect is inconceivable, even for those accounts (such as Evans's recent work) which do consider the possibility of cognitive biases also arising in type 2 processes. True enough, on some accounts (Gigerenzer 2007), fast-and-frugal type 1 processes are viewed as systematically *outperforming* type 2 processes, but not in virtue of bringing performance closer to normative standards as dictated by the deductive canons or some other theoretic normative model. The claim is that fast-and-frugal heuristics *themselves* would outperform these theoretical normative models. So the present account of the debiasing effect of formal languages does not fit into those conceptualizations which emphasize the power of fast-and-frugal type 1 processes either, in particular in that it is claimed that operating with formal languages has a truly mind-altering, debiasing effect (section 5.2.3).

---

[27] Although it seems that not all proponents of dual-process theories would endorse the view that all type 1 processes are non-verbal.

I hasten to add that I am not rejecting all of the basic insights of dual-process theories: ultimately, a plethora of evidence (e.g., work on implicit biases in social psychology) clearly supports the idea of distinct, parallel processes in human cognition, but the main distinction seems to be between conscious v. pre-conscious, and to some extent between controlled v. automatic. The main problem with most dual-process theories lies in how the dichotomy is characterized, which seems to rely mostly on old dogmas such as the oppositions between perception and reason, lower cognition v. higher cognition, and so forth.[28] In particular, on the view of human reasoning defended in section 4.3, 'logical thinking' (if defined according to the classical deductive canons) is neither a type 1 process nor a type 2 process; it simply is not there to start with, as it must be learned/trained for to be mastered, and the learning process will engage both so-called type 1 and type 2 processes.

The 'logic' that does seem to underpin human cognition generally speaking, *au naturel*, so to speak, essentially follows non-monotonic patterns, as argued by Stenning and van Lambalgen and discussed in Chapter 4.[29] From the point of view of a non-monotonic logic of planning, there may not be significant differences between perception, motor planning, and so-called 'higher cognition':

We plan with respect to our expectations of the world, not, as we would have to if we planned using classical logic, with respect to all logical possibilities. Maintaining a model of the current state of the immediate environment relevant to action is a primitive biological function; calculating what is true in all logically possible models of the current sense data is not. These planning logics are just as much what one needs for planning low-level motor actions such as reaching and grasping, as they are for planning chess moves. (Stenning and van Lambalgen 2008: 144)

---

[28] Keren and Schul (2009) provide a critique of the conceptual underpinning of dual-process theories. Here is a passage from the abstract: 'We conclude that two-system models currently provide little scientific advance, and we encourage researchers to adopt more rigorous conceptual definitions and employ more stringent criteria for testing the empirical evidence in support for two-system theories.'

[29] Stanovich, who coined the concept of the 'fundamental computational bias' in human cognition (see section 4.2, above), explicitly relates it to a dual-system framework. 'Stanovich also addresses questions of evolution and rationality. He argues that System 1 was designed for the promotion of narrowly genetic goals, such as reproductive success, whereas the more flexible System 2 serves the goals of the individual person and allows us to rebel against genetic imperatives (it is still an evolutionary product, of course, but is under "long-leash" genetic control). In modern technological and bureaucratic societies, Stanovich argues, success often requires us to engage in abstract, decontextualised reasoning and to suppress System 1 processes, with their tendency to contextualise problems (the "fundamental cognitive bias")' (Frankish 2010: 920). The point here is that while Stanovich's description of the fundamental computational bias seems essentially on target it is far from obvious that the bias 'inhabits' System 1 and is compensated for exclusively by System 2; it seems to inhabit both 'systems', and can be compensated for in different ways.

Moreover, it is typically held that type 1 processes are characterized by slow acquisition and change – in other words, that learning and training have little effect over them, and would mostly 'belong' to the type 2 realm. Now, insofar as sensorimotor cognitive processes belong to the type 1 realm, this seems to be at odds with the fact that we do learn to master a wide range of sensorimotor cognitive processing very proficiently, which then become largely automatic – driving, for example. More generally, the (often tacit) presupposition that automaticity and learning/acquisition are mutually exclusive concepts is simply false, as suggested, for example, by the high level of automaticity in semantic activation upon reading, which is obviously a learned skill. Similarly, a proficient user of a given formalism typically operates with it in a semi-automatic rather than a controlled manner, even though the learning process typically requires a much more conscious approach.

In other words, for a large chunk of our valuable cognitive skills, the goal is precisely to go, through learning, from conscious, controlled processing to a certain level of automaticity and non-consciousness, which is precisely what characterizes expertise. This holds in particular of cognitive activities involving sensorimotor processing, but is arguably a very general phenomenon. In sum, the type 1 v. type 2 distinction as usually characterized is inadequate to make sense of a wide range of learning processes, in particular in that it seems to view learning as mostly connected to type 2 processing.

This is not the place for a thorough investigation of the conceptual shortcomings of the dual-process theories available in the market; here, I just want to suggest that the failure of the present account of the debiasing effect of formal languages (in terms of de-semantification and sensorimotor processing) to fit into the standard dual-process approaches should not be taken as evidence against the account. Rather, this suggests that the type 1 v. type 2 dichotomy as currently conceptualized is too crude to be able to make sense of a wide range of interesting cognitive phenomena. In particular, debiasing need not be only a matter of higher-cognition, 'type-2' processes; debiasing can come about also through a certain level of automaticity and sensorimotor processing. In fact, while I do not dispute that appeal to meta-cognition can have a debiasing effect, in practice it is often (though not always) the case that the more we think about what we are doing the more prone we are to biased reasoning.[30]

---

[30] Studies in post-rationalization seem to confirm this claim (cf. Mercier and Sperber 2011: References). This being said, awareness of one's own biases can certainly facilitate the process of counterbalancing for them in situations where this might be advantageous (e.g., decision-making, medical diagnosis, counterbalancing doxastic conservativeness in scientific practice, etc.).

7.3 CONCLUSION

This chapter began with a philosophical, schematic discussion of the very concept of formalization; this was a necessary step towards the explanation of how de-semantification and sensorimotor manipulation of the notation are combined for the debiasing effect of reasoning with formal languages and formalisms to come about.

In the second part, a very different conception of debiasing was discussed, namely the conception investigated by Houdé and collaborators, which (*pace* their own conceptualization of it) seems to rely essentially on meta-cognition and cognitive monitoring. This conception of debiasing prescribes that one should think 'harder' to reason logically, whereas the debiasing effect of formal languages as presented here consists precisely in recommending that the agent think 'less' about what she is doing and let the formalism 'do the thinking'; it is based on the ideas of *externalization* and *automatization* (mechanization) of reasoning processes. Now, while the first conception of debiasing falls squarely within the framework of dual-process theories of cognition, the second conception goes in the opposite direction in that it seems to attribute a debiasing capacity to *type 1* processes (automatic, perceptual/sensorimotor).

The claim is not that this conception of debiasing is the only correct/ effective one; there is no doubt that cognitive monitoring also seems to offer great debiasing potential. Rather, the claim is that, if I am right in my description of the debiasing effect of formalisms – which is moreover firmly grounded in a distinguished tradition, ranging from Lull and Leibniz to Turing, of emphasis on 'mechanical' logical reasoning – then dual-process theories simply do not seem to be able to account for this significant cognitive phenomenon.

At any rate, the goal of this final chapter was to clarify once and for all the main thesis of the book, namely that reasoning with formal languages and formalisms can have a truly mind-altering effect; it may allow human agents to counter some of their most deeply engrained cognitive biases, such as the tendency to rely on and seek to 'hold on' to prior belief. Formal languages and formalisms offer a unique cognitive boost to human agents.

Conclusion

The present investigation is clearly an investigation on *methodology*, more specifically on the ins and outs of *formal* methods – in logic, philosophy, and elsewhere. Perhaps ironically, formal methods are extensively discussed but not in fact employed here; this is an investigation on logical formulae with very few logical formulae. Instead, the methodology adopted is what I like to describe as *integrative*, combining 'traditional' philosophical analysis (whatever that means – see below) with attention to both empirical and historical elements. While philosophical analysis and empirical elements are increasingly often combined in mainstream philosophy (yielding what is often referred to as 'empirically informed philosophy'), and while the role of historical analysis for philosophical theorizing has long (albeit not unanimously) been recognized, the combination of these three approaches is thus far rather unusual. One exception might be Netz's (1999) project of 'cognitive history', which nevertheless for now remains essentially at a programmatic level.

Now, as stated in the introduction, I believe that a wide range of philosophical questions can only be properly treated once these three approaches are combined. At any rate, the questions motivating *this* investigation are among those which (arguably) require the integrative perspective. The main questions pertain to the import and impact of formal languages on the cognitive practices of human users, i.e., what it means for a reasoner to reason aided by these specific tools. So, naturally, the investigation had to rely extensively on empirical data about human cognition, both assisted and unassisted by the cognitive scaffolding offered by formal languages and formalisms. Moreover, it was imperative to emphasize the long historical development leading to the current technological stage of formal languages, so as to challenge the apparent obviousness of doing logic (or physics, or economics, etc.) with these particular tools. But given the somewhat idiosyncratic nature of the integrative approach, it seems that a more sustained discussion of the methodology adopted is still called for, and this will be the main focus of this concluding chapter.

DIFFERENT METHODOLOGIES IN PHILOSOPHICAL INVESTIGATION

Looking at past and present work in philosophy, it would seem that we can identify four important strands of methodological approaches in philosophical analysis.[1] This proposed taxonomy is likely not to be exhaustive, but it attempts to be encompassing. Moreover, as it is based on a bottom-up approach, no claim is made to represent the 'essence' or 'core' of philosophical theorizing. In other words, the (future) appearance of new methodologies for philosophical theorizing is not to be ruled out.

With these caveats in place, the four proposed categories are:

(i) Conceptual, a priori reflection

This category would correspond to what I have referred to above (rather imprecisely) as 'traditional philosophical methods'. This is a rather heterogeneous category, and thus difficult to characterize in just a few lines; but the common denominator is that these methods can be applied 'from the armchair', as inner reflection – the paradigmatic example being Descartes 'clothed in my dressing gown, seated next to the fireplace'. Conceptual analysis in particular is a core component, as philosophy remains under the influence of the Socratic enterprise of unpacking the precise meaning of familiar concepts ('What is time? What is knowledge?'). Besides conceptual analysis, the production of novel concepts is another important component, as is reasoning more or less 'deductively' (drawing conclusions that follow necessarily from certain premises) about a given subject-matter.

In analytic philosophy in particular, a common approach to conceptual analysis consists in formulating necessary and sufficient conditions for something to count as falling under concept A. We could say that (analytic) philosophers often seek precisifications of the following schema:

$$x \text{ is } A \Leftrightarrow x \text{ is } B, C, D \ldots$$

where A is the concept to be analysed in terms of (presumably) simpler concepts B, C, D... A classical example is the definition of the concept of *knowledge* as *justified true belief*. But, given a proposed/accepted definition,

---

[1] There may well be other categories not discussed here, but, for the purposes of the present analysis, these are the four key categories.

philosophers then set out to look for counterexamples, i.e., instances of *x* which either satisfy the definiens but not the definiendum, or the other way round. Here, the classical examples are Gettier cases, which appear to be instances of justified true belief, and yet arguably (i.e., according to Gettier and the tradition following him) do not count as knowledge. Thus, a cornerstone of this approach is its reliance on so-called 'pre-theoretical intuitions', corresponding to the extra-theoretical judgment that *x* is indeed A even though it fails to be an instance of B, C, D, . . . , or vice versa. These intuitions are often treated as having ultimate epistemic authority, and, once a counterexample of this kind is found, philosophers then attempt to reformulate the definition of A so as to accommodate the purported counterexample.

A wide range of other methodological principles and devices belong in this category (e.g., philosophical thought-experiments), but the general idea is that this kind of theorizing requires no recourse to 'external' sources (in particular, external empirical knowledge); all the necessary material is available to the thinker by introspection and reflection alone.

(ii) Formal methods

Formal methods in philosophy correspond to the application of mathematical and logical tools to the investigation of philosophical issues. As examples, one could cite the development of possible world semantics for the analysis of the concepts of necessity and possibility, applications of the Bayesian framework to issues in epistemology (giving rise to so-called formal epistemology), Carnapian explication, and many others.

In a sense, insofar as logic underpins Aristotle's philosophical investigations across the board (which is in fact a moot point), and insofar as logic in general provided the overall methodology for philosophical investigation in the Latin medieval tradition, 'formal' methods have a long and distinguished history in philosophical practice. But, when speaking of formal methods in philosophy, it is specifically the application of *symbolic, mathematical logic* to philosophical problems that one has in mind, and this only started with Russell (possibly with Frege). Naturally, formal methods share the a priori nature of the previous category, but the crucial difference is the 'language' in which the investigation is conducted: semi-regimented but essentially 'everyday' languages in the previous category v. the 'formal' languages of logic and mathematics. As I have argued throughout the book, this difference makes all the difference.

(iii) Historical methods

What I refer to as 'historical methods' in philosophical theorizing consists essentially in tracing the historical origins of a given philosophical concept or practice in order to attain a better understanding of its *current* instantiations. The enterprise of unearthing where philosophical concepts, theses, and dogmas come from can be described as 'conceptual archaeology'.

Naturally, the study of the history of philosophy is an autonomous enterprise as such; moreover, much interesting philosophical theorizing emerges from 'purely historical' investigations. But here the point is the potential usefulness of history for the systematic investigation of a currently prominent philosophical concept or thesis. An important difference with respect to how history of philosophy is typically studied is the focus on concepts rather than on authors: conceptual archaeology produces stories 'whose protagonists are not people, but concepts, problems, rules or arguments'.[2]

The importance of contextualizing the discussion of a given topic historically even if the investigative goals are essentially non-historical is a noteworthy component of the so-called 'continental' tradition in philosophy, broadly construed. The concepts of genealogy (Nietzsche) and archaeology (Foucault) are some of the instantiations of the importance attributed to historical investigation for philosophical theorizing within this tradition. Analytic philosophers who are not specifically historians of philosophy, by contrast, often question the relevance of history for philosophical theorizing: the underlying assumption seems to be that concepts are essences ('natural kinds'), and from this point of view the chronological unfolding of concepts is a merely historical, philosophically unimportant contingency. In contrast, continental philosophers will (generally speaking) tend to view concepts as historically construed entities, and from this point of view an analysis of a concept cannot be dissociated from an analysis of its history.

Beyond this admittedly oversimplified picture of the 'analytic v. continental' distinction, the main point for now is to characterize the notion of historical methods in philosophy as *distinct* from the study of the history of philosophy as a goal in itself (naturally, the former can nevertheless greatly benefit from the latter). In the next section, reasons to deploy historical methods thus understood will be discussed in more detail.

---

[2] In the apposite terms of Alain de Libera, in an interview dated 4 January 2009. www.actu-philosophia. com/spip.php?article77. De Libera's conception of 'archeology' as a philosophical enterprise is very similar to the conception proposed here.

(iv) Empirically informed methods

These are the approaches characterized by the systematic reference to data and results from the empirical and social sciences, such as the sciences of the mind, physics, biology, sociology, anthropology, etc. Sometimes these approaches are described as 'naturalistic', but, as is well known, there are (too?) many variations of the concept of naturalism in philosophy (Papineau 2007); I submit that 'empirically informed' is a more precise label.

In recent years, some philosophers took up the challenge of conducting empirical research themselves, in what came to be known as the *experimental philosophy* movement. Broadly speaking, experimental philosophy (X-phi) belongs under the heading of empirically informed methods, but it is important to stress that this is just one specific approach in a larger group of empirically informed approaches. Given the methodological intricacies of conducting empirical research, many philosophers prefer to leave this aspect of the investigation to those with the specific training for it, and to concentrate on the interpretation of the potential philosophical significance of the results.

But philosophers need not take the interpretations of the results by scientists or even the results themselves at face value; as illustrated in Chapter 4, above, it is perfectly possible for philosophers to engage critically with the empirical material (in particular, questioning hidden assumptions and presuppositions), while at the same time recognizing its value for philosophical theorizing. In effect, it is often possible to 'filter' these analyses so as to separate conceptual confusion from valuable empirical results.

For most of its history, philosophy as an intellectual enterprise was not viewed as sharply demarcated and isolated from other sciences and fields of inquiry, in particular the empirical sciences. Aristotle was above all a biologist, and much of his metaphysics and ethics arguably emerges from his biological investigations; Descartes and Leibniz were mathematicians and physicists just as much as philosophers. I submit that it is only with Kant's critical project that the emphasis on sharply demarcating philosophy from other disciplines acquired prominent status. Thus, the recent re-emergence of interdisciplinary investigations involving philosophy and the empirical and social sciences (very prominent in, for example, current philosophy of mind, in particular since the publication of Churchland's ground-breaking *Neurophilosophy* (1986)) can in fact be viewed as a return to a pre-Kantian, scientifically informed conception of philosophy.

WHAT EACH OF THEM HAS TO OFFER

As is perhaps already apparent from the previous section, I claim that each of these methods has something different and crucial to offer to philosophical theorizing. The first category, 'traditional' philosophical methods, contributes with the clarification of key concepts for the investigation and the production of new concepts. Essentially, it will ensure that the conceptual foundations of a given theory or investigation are sufficiently firm and solid. As for formal methods, as has been argued throughout the book, they not only add precision and clarity, but also seem to allow for the derivation of non-trivial conclusions from assumptions, which would be difficult or perhaps even impossible to attain with purely conceptual means.

Historical methods can yield the benefic result of bringing to the fore the origins of our concepts and 'philosophical intuitions', and of isolating the (possibly contentious) assumptions and substantive theoretical steps that they depend on. They outline the contingency of much of what we take for granted, and, insofar as philosophy is to a large extent the enterprise of questioning the obvious, historical methods are irreplaceable in order to maintain a critical stance towards what is taken to be *philosophically* obvious by philosophers themselves.

Empirical methods help us keep our philosophical confabulations 'in check'; they may increase the explanatory value of philosophical theories, taking them beyond the status of 'just-so stories'. Insofar as philosophy purports to offer accounts not only of *possible* but also of *actual* phenomena, it simply cannot ignore the picture of the world produced by our best current sciences (which, however, does not entail a conception of science as infallible). On conceptual grounds alone, i.e., relying on criteria such as coherence and generality, competing philosophical theories will often come out as equally 'good', but only those that are at least consistent with the available empirical data can claim to be adequate theories of phenomenon X. This said, given a logical space of different accounts for X, what empirical data are likely to do is to narrow down the theories still 'on the race' so to speak, but not necessarily point in the direction of the unique correct theory (as in fact is often the case in the empirical and social sciences as well). But, at the very least, philosophy must remain 'empirically responsible'.[3]

As may have become apparent by now, the three methodologies – formal, historical, empirical – also offer the possibility of counterbalance to the potentially exaggerated role of 'intuitions' in philosophical investigations.

---

[3] I owe this lovely phrase to John Protevi.

Formal methods can help counter excessive doxastic conservativeness, as argued throughout the book; historical methods emphasize the contingency of many of our philosophical intuitions by revealing where they come from; empirical methods may show that many of our deeply engrained 'intuitions' are simply incorrect accounts of certain phenomena (e.g., most of what is often described as 'folk physics'). More generally, it is quite likely that the different methodologies complement and counterbalance each other in a variety of ways, which for reasons of space will not be further explored at this point.

HOW THEY CAN BE COMBINED

If the different methodological approaches have different but complementary theoretical benefits to offer, a natural conclusion to be drawn is that it is advisable to *combine* them for philosophical theorizing. Indeed, I submit that the most reasonable position with respect to philosophical methodology is what I call *conjunctive methodological pluralism*: the idea that not only is there room for distinct methodological approaches in philosophy but also that they should be *combined* in one and the same investigation. While I do not rule out that some philosophical questions may still be most adequately treated by means of a uniform methodology (in particular, conceptual, a priori reflection), I submit that a very wide range of philosophical questions in fact require the integration of different approaches.

Naturally, the obvious question to be asked as this point is: but *how*? How does one combine methods that are so dissimilar and which rely on different desiderata and principles? This is undoubtedly a real concern, which for reasons of space will only be briefly discussed here. Nevertheless, what seems to me to be the key component of a successful integrative investigation is that it be based on one underlying *core hypothesis* which translates into research questions and predictions in each methodological sphere.

The core hypothesis of the present investigation is that formal languages are best seen as cognitive technologies which not only augment but truly *alter* the cognitive and reasoning capacities of human agents. This hypothesis can be readily translated into historical as well as empirical research questions and predictions: the historical development of formal languages can be viewed as a gradual search for cognitively more powerful notational tools; the actual cognitive impact of using formal languages and formalisms for reasoning can be discussed on the basis of empirical results from research in psychology and cognitive science. The hypothesis also had important purely philosophical/conceptual implications, which have been addressed as well.

This said, it would be a mistake to conclude that philosophical theorizing should be *reduced* to the empirical/social sciences. While there may be much overlap (contrary to Kantian demarcational ideals), philosophy remains an enterprise with its own characteristic flavour. Philosophers are well placed to undertake the kind of conceptual, foundational analysis that is required for any investigative enterprise. Indeed, this is where the collaboration between philosophy and the empirical/social sciences may be viewed as a two-way street after all: not only do we philosophers benefit from the empirical results from these sciences, but scientists can benefit from our skills of conceptual clarity and analysis.[4]

THE DIFFERENT METHODOLOGIES IN THE PRESENT WORK

As must be clear by now, the present analysis is entirely premised on the notion of conjunctive methodological pluralism; in particular, conceptual, historical, and empirically informed approaches have been combined to discuss the status of formal languages as cognitive technologies. (Some formal results have been mentioned briefly in Chapter 3, in particular concerning the phenomena of incompleteness, non-categoricity, and the orthogonal desiderata of expressiveness and tractability, but it would be incorrect to say that formal methods have been extensively deployed here.)

The general background and argumentative strategies (as well as the vocabulary and linguistic style) adopted fall undoubtedly for the most part under the 'genre' of analytic philosophy, even though conceptual analysis in terms of necessary and sufficient conditions has often been supplanted by analyses based on the concept of prototypical features (as, for example, the concept of language in Chapter 2). Still, it would seem that there are at least a few purely (and essentially 'analytic') conceptual sections scattered in the book (in particular, much of Chapter 2, some of Chapters 1 and 3, and section 7.1).

The presence of historical methods may broadly be seen as falling under the influence of the 'continental' notions of genealogy and archaeology. As part of the present research project, I have undertaken an 'archaeological' investigation of the notion of the formal with respect to logic (cf. especially

---

[4] This does not imply that the *only* relevance of philosophy is to 'serve' the empirical sciences. But, as anecdotal evidence, I can report that, in my recent dealings with psychologists and cognitive scientists, some of them (Keith Stenning, David Over, and Shira Elqayam, in particular) have been quite interested in the kind of philosophical/conceptual 'deconstruction' of some of the theoretical pillars of their research that I can offer. For example, I discuss the philosophical background of normativism in psychology in my commentary (Dutilh Novaes 2011d) on Elqayam and Evans 2011.

Dutilh Novaes 2011b; 2012). From this investigation, two senses of the formal have emerged as particularly relevant for the concept of a formal language and for the conceptualization of a formal language as a cognitive technology.

Moreover, it seemed crucial to describe formal languages not as readily available tools but rather as a technology having developed quite gradually, as was done in Chapter 3. The main theses of the book could simply not have been argued for satisfactorily without at least a brief account of the historical development of formal languages.

Finally, the ubiquity of references to empirical results from different scientific domains (especially in Chapters 4, 5, 6, and 7, but also in Chapter 2 when discussing the concept of a language) is perhaps the main differential of the present investigation. I have incorporated data from different sub-fields of psychology and cognitive science, such as research on (deductive) reasoning (Chapter 4), the cognitive impact of working with external symbolic systems (Chapter 5), semantic activation (Chapter 6), and inhibition and dual-process theories of cognition (Chapter 7). Interestingly, these different sub-fields are often not conversant with one another (it is particularly striking that no connection seems to have been made in the literature between the phenomenon of matching bias and results on semantic activation), but each seems to contribute to the elucidation of the main research questions raised here in different ways.

I hope that the present work will be seen not only as a contribution to our understanding of the impact of reasoning with formal languages and formalisms but also as a vindication of the importance of historical and empirically informed approaches for philosophical theorizing. I also hope to have shown that the different methodologies can (and often should!) be combined for the discussion of central philosophical issues.

Let me close with a personal anecdote. On one occasion when presenting this material someone (a good friend of mine, as a matter of fact) raised the following 'objection': 'Look, this is all very interesting, but that's not philosophy.' I must confess that I was not very moved by this objection; I do not see it as a real problem if this investigation does not conform to a given, perhaps overly narrow, conception of what philosophy is or should be. I would have been much more concerned if the objection had been, 'Look, this is all very philosophical, but it's not very interesting.' (So far, this objection at least has not been raised in my presence.)

References

Adams, E. W. 1965. 'A Logic of Conditionals'. *Inquiry* 8: 166–97.

1975. *The Logic of Conditionals*. Dordrecht: Reidel.

Adams, F., and K. Aizawa. 2001. 'The Bounds of Cognition'. *Philosophical Psychology* 14: 43–64.

2008. *The Bounds of Cognition*. Oxford: Blackwell.

Albert, M. L., A. Yamadori, H. Gardner, and D. Howes. 1973. 'Comprehension in Alexia'. *Brain* 96.

Amodio, D. M., J. T. Kubota, E. Harmon-Jones, and P. G. Devine. 2006. 'Alternative Mechanisms for Regulating Racial Responses According to Internal vs External Cues'. *Social Cognitive and Affective Neuroscience* 1: 26–36.

Anderson, S. W., A. R. Damasio, and H. Damasio. 1990. 'Troubled Letters but Not Numbers'. *Brain* 113.

Andrade-Lotero, E., and C. Dutilh Novaes. 2012. 'Validity, the Squeezing Argument and Alternative Semantic Systems: The Case of Aristotelian Syllogistic'. *Journal of Philosophical Logic* 41: 387–418.

Andrews, G. 2010. 'Belief-based and Analytic Processing in Transitive Inference Depends on Premise Integration Difficulty'. *Memory and Cognition* 38: 928–40.

Arianrhod, R. 2003. *Einstein's Heroes*. Cambridge: Icon Books.

Aristotle. 1984. 'Physics'. In *The Complete Works of Aristotle*, ed. J. Barnes. Vol. 1. Princeton University Press.

Awodey, S., and E. Reck. 2002. 'Completeness and Categoricity, Part I: Nineteenth-century Axiomatics to Twentieth-century Metalogic'. *History and Philosophy of Logic* 23: 1–30.

Barker-Plummer, D. 1995. 'Turing Machines'. Ed. E. Zalta. *Stanford Encyclopedia of Philosophy*. http://plato.stanford.edu/entries/turing-machine/.

Barsalou, L. W. 1999. 'Perceptual Symbol Systems'. *Behavioral and Brain Sciences* 22(4): 577–609.

Baumgartner, M., and T. Lampert. 2008. 'Adequate Formalization'. *Synthese* 164(1): 93–115.

Bechtel, W. 1994. 'Natural Deduction in Connectionist Systems'. *Synthese* 101: 433–63.

Bellos, A. 2010. *Alex's Adventures in Numberland*. London: Bloomsbury.

Benacerraf, P. 1973. 'Mathematical Truth'. *Journal of Philosophy* 70(19): 661–79.

Bensaude-Vincent, B., and W. R. Newman 2007. 'Introduction to the Artificial and the Natural: State of the Problem'. In B. Bensaude-Vincent and W. R. Newman (eds.), *The Artificial and the Natural: An Evolving Polarity.* Cambridge, MA: MIT Press.

van Benthem, J. 2011. 'The Dynamic World of Martin Stokhof'. In C. Dutilh Novaes and J. van der Does (eds.), *Festschrift for Martin Stokhof.* Available at www.vddoes.net/Martin/mf.html.

Bernays, P. 1930. 'The Philosophy of Mathematics and Hilbert's Proof Theory'. In P. Mancosu (ed.), *From Brouwer to Hilbert.* Oxford University Press: 234–65.

Blanchette, P. 2007. 'The Frege–Hilbert Controversy'. Ed. E. Zalta. *Stanford Encyclopedia of Philosophy.* http://plato.stanford.edu/entries/frege-hilbert/.

Bobzien, S. 2006. 'Ancient Logic'. *Stanford Encyclopedia of Philosophy.* http://plato.stanford.edu/entries/logic-ancient/.

Boole, G. 1847. *The Mathematical Analysis of Logic, Being an Essay Towards a Calculus of Deductive Reasoning.* Cambridge: Macmillan, Barclay, and Macmillan.

　1854. *An Investigation of the Laws of Thought on which are Founded the Mathematical Theories of Logic and Probabilities.* London: Macmillan.

Braine, M. D. S., and D. P. O'Brien. 1998. *Mental Logic.* London: Routledge.

Brandom, R. 1994. *Making it Explicit: Reasoning, Representing, and Discursive Commitment.* Cambridge, MA: Harvard University Press.

Brun, G. 2003. *Die richtige Formel: Philosophische Probleme der logischen Formalisierung.* Frankfurt/Main: Ontos.

Burris, S. 2009. 'The Algebra of Logic Tradition'. Ed. E. Zalta. *Stanford Encyclopedia of Philosophy.* http://plato.stanford.edu/entries/algebra-logic-tradition/.

　2010. 'George Boole'. Ed. E. Zalta. *Stanford Encyclopedia of Philosophy.* http://plato.stanford.edu/entries/boole/.

Byrne, R. M. J. 1989. 'Suppressing Valid Inferences with Conditionals'. *Cognition* 31: 1–83.

Carnap, R. 1934. *The Logical Syntax of Language.* London: Open Court.

Carruthers, P. 2002. 'The Roots of Scientific Reasoning: Infancy, Modularity and the Art of Tracking'. In P. Carruthers, S. Stich, and M. Siegal (eds.), *The Cognitive Basis of Science.* Cambridge University Press: 73–95.

Castro-Caldas, A., K. M. Peterson, A. Reis, S. Stone-Elander, and M. Ingvar. 1998. 'The Illiterate Brain: Learning to Read and Write during Childhood Influences the Functional Organisation of the Adult Brain'. *Brain* 121: 1053–63.

Chemero, A., and M. Silberstein. 2008. 'Defending Extended Cognition'. In V. Sloutsky, K. McRae, and B. C. Love (eds.), *Proceedings of the 30th Annual Meeting of the Cognitive Science Society, Washington.* Austin, TX: Cognitive Science Society: 129–34.

Chemla, K. 2005. 'The Interplay between Proof and Algorithm in 3rd Century China: The Operation as Prescription of Computation and the Operation as

Argument'. In P. Mancosu (ed.), *Visualization, Explanation and Reasoning Styles in Mathematics*. Synthese Library. Berlin: Springer Verlag.

Cheng, P., and K. Holyoak. 1985. 'Pragmatic Reasoning Schemas'. *Cognitive Psychology* 14.

Chevallier, C., I. A. Noveck, T. Nazir, L. Bott, V. Lanzetti, and D. Sperber. 2008. 'Making Disjunctions Exclusive'. *Quarterly Journal of Experimental Psychology* 61(11): 1741–60.

Church, Alonzo. 1936. 'A Note on the Entscheidungsproblem'. *Journal of Symbolic Logic* 1: 40–1.

Churchland, P. 1986. *Neurophilosophy: Toward a Unified Science of the Mind–Brain*. Cambridge, MA: MIT Press.

Cipolotti, L., and B. Butterworth. 1995. 'Towards a Multiroute Model of Number Processing: Impaired Number Transcoding and Preserved Calculation Skills'. *Journal of Experimental Psychology: General* 124: 375–90.

Cipolotti, L., B. Butterworth, and G. Denes. 1991. 'A Specific Deficit for Numbers in Case of Dense Acalculia'. *Brain* 114: 2619–37.

Cipolotti, L., E. Warrington, and B. Butterworth. 1995. 'Selective Impairment in Manipulating Arabic Numerals'. *Cortex* 31: 73–86.

Clark, A. 2003. *Natural-born Cyborgs: Minds, Technologies, and the Future of Human Intelligence*. Oxford University Press.

2008. *Supersizing the Mind*. Oxford University Press.

Clark, A., and D. Chalmers. 1998. 'The Extended Mind'. *Analysis* 58: 10–23.

Cook, R. 2010. 'Let a Thousand Flowers Bloom: A Tour of Logical Pluralism'. *Philosophy Compass* 5(6): 492–504.

Copeland, B. J. 1997. 'The Church–Turing Thesis'. Ed. E. Zalta. *Stanford Encyclopedia of Philosophy*. http://plato.stanford.edu/entries/church-turing/.

Corcoran, J. 2003. 'Aristotle's *Prior Analytics* and Boole's *Laws of Thought*'. *History and Philosophy of Logic* 24: 261–88.

Cosmides, L. 1989. 'The Logic of Social Exchange: Has Natural Selection Shaped How Humans Reason?'. *Studies with the Wason Selection Task* 31: 187–276.

Counihan, M. 2008a. 'Looking for Logic in all the Wrong Places: An Investigation of Language, Literacy and Logic in Reasoning'. PhD dissertation. University of Amsterdam.

2008b. 'If p then q ... and all that: Logical Elements in Reasoning and Discourse'. *Journal of Logic, Language and Information* 17(4): 391–415.

De Cruz, H. 2007. 'How Does Complex Mathematical Theory Arise? Phylogenetic and Cultural Origins of Algebra'. In C. Gershenson, D. Aerts, and B. Edmonds (eds.), *Worldviews, Science and Us: Philosophy and Complexity*. Hackensack, NJ: World Scientific: 338–51.

De Cruz, H., and J. De Smedt. 2010. 'The Innateness Hypothesis and Mathematical Concepts'. *Topoi* 29: 3–13.

De Cruz, H., H. Neth, and D. Schlimm. 2010. 'The Cognitive Basis of Arithmetic'. In B. Löwe and T. Müller (eds.), *Philosophy of Mathematics: Sociological Aspects and Mathematical Practice*. London: College Publications: 39–86.

Curry, H. B. 1957. *A Theory of Formal Deducibility*. University of Notre Dame Press.

Damasio, A. 1994. *Descartes' Error: Emotion, Reason, and the Human Brain*. New York: Putnam Publishing.

Dedekind, R. 1890 [1981]. 'Letter to Keferstein'. In J. van Heijenoort (ed.), *From Frege to Gödel: A Source Book in Mathematical Logic, 1879–1931*. Cambridge, MA: Harvard University Press: 98–103.

Dehaene, S. 2005. 'Evolution of Human Cortical Circuits for Reading and Arithmetic: The "Neuronal Recycling" Hypothesis'. In J. Duhamel, G. Rizzolatti, S. Dehaene, and M. Hauser (eds.), *From Monkey Brain to Human Brain*. Cambridge, MA: MIT Press: 133–57.

2009. *Reading in the Brain: The Science and Evolution of a Human Invention*. New York: Viking.

Dehaene, S., F. Pegado, L. Braga, P. Ventura, G. Filho, A. Jobert, G. Dehaene-Lambertz, R. Kolinsky, J. Morais, and L. Cohen. 2010. 'How Learning to Read Changes the Cortical Networks for Vision and Language'. *Science* 330(6009): 1359–64.

Demichelis, S., and J. W. Weibull. 2008. 'Language, Meaning, and Games: A Model of Communication, Coordination, and Evolution'. *American Economic Review* 98(4): 1292–311.

Dennett, D. C. 1995. *Darwin's Dangerous Idea: Evolution and the Meanings of Life*. Harmondsworth: Penguin.

Derrida, J. 1998. *Of Grammatology*. Baltimore, MD: Johns Hopkins University Press.

Descartes, R. 1985. *The Philosophical Writings of Descartes*. Ed. J. Cottingham, R. Stoothoff, and D. Murdoch. Vol. 1. Cambridge University Press.

Donald, M. 1991. *Origins of the Modern Mind*. Cambridge, MA: Harvard University Press.

2001. *A Mind So Rare: The Evolution of Human Consciousness*. New York: W. W. Norton.

Dunbar, K. N. 2002. 'Understanding the Role of Cognition in Science: The Science as Category Framework'. In P. Carruthers, S. Stich, and M. Siegal (eds.), *The Cognitive Basis of Science*. Cambridge University Press: 154–70.

Dutilh Novaes, C. 2004. 'A Medieval Reformulation of the de Dicto / de Re Distinction'. *LOGICA Yearbook 2003, Prague, Filosofia, 2004*: 111–24.

2007. *Formalizing Medieval Logical Theories: Suppositio, Obligationes and Consequentia*. Berlin: Springer.

2008. '14th Century Logic after Ockham'. In D. Gabbay and J. Woods (eds.), *The Handbook of the History of Logic*. Vol. 2. Amsterdam: Elsevier.

2010a. 'Surprises in Logic'. In M. Palis (ed.), *LOGICA Yearbook 2009*. London: College Publications.

2010b. ''He doesn't want to prove this or that': On the Very Young Wittgenstein'. *Philosophical Books* 51(2): 102–16.

2011a. 'Medieval Theories of Supposition'. Ed. H. Lagerlund. *Encyclopedia of Medieval Philosophy: Philosophy between 500 and 1500*. Berlin: Springer Verlag.

2011b. 'The Different Ways in which Logic is (Said to Be) Formal'. *History and Philosophy of Logic* 32: 303–32.

2011c. 'Lessons on Truth from Medieval Solutions to the Liar Paradox'. *Philosophical Quarterly* 61: 58–78.

2011d. 'The Historical and Philosophical Origins of Normativism'. *Behavioral and Brain Sciences* 34: 253–4.

2012. 'Reassessing Logical Hylomorphism and the Demarcation of Logical Constants'. *Synthese* 185: 387–41.

forthcoming. 'Mathematical Reasoning and External Symbolic Systems'. *Logique and Analyse*.

Edgington, D. 1995. 'On Conditionals'. *Mind* 104: 235–329.

2001. 'Conditionals'. Ed. E. Zalta. *Stanford Encyclopedia of Philosophy*. http://plato.stanford.edu/entries/conditionals/.

Elio, R. 2002. *Common Sense, Reasoning, and Rationality*. Oxford University Press.

Elqayam, S., and J. Evans. 2011. 'Subtracting Ought from Is: Descriptivism versus Normativism in the Study of Human Thinking'. *Behavioral and Brain Sciences* 34(5): 233–48.

Etchemendy, J. 2008. 'Reflections on Consequence'. In D. Patterson (ed.), *New Essays on Tarski and Philosophy*. Stanford, CA: CSLI Publications: 263–99.

Evans, J. 1998. 'Matching Bias in Conditional Reasoning: Do We Understand it after 25 Years?' *Thinking and Reasoning* 4: 45–110.

2002. 'Logic and Human Reasoning: An Assessment of the Deduction Paradigm'. *Psychological Bulletin* 128(6): 978–96.

2006. 'The Heuristic–Analytic Theory of Reasoning: Extension and Evaluation'. *Psychonomic Bulletin and Review* 13(3): 378–95.

2007. *Hypothetical Thinking: Dual Processes in Reasoning and Judgement*. Hove: Psychology Press.

Evans, J., J. Barston, and P. Pollard. 1983. 'On the Conflict between Logic and Belief in Syllogistic Reasoning'. *Memory and Cognition* 11: 295–306.

Evans, J., and K. Frankish (eds.). 2009. *In Two Minds: Dual Processes and Beyond*. Oxford University Press.

Evans, J., S. E. Newstead, J. L. Allen, and P. Pollard. 1994. 'Debiasing by Instruction: The Case of Belief-bias'. *European Journal of Cognitive Psychology* 6(3): 263–85.

Evans, J., and D. Over. 2004. *If*. Oxford University Press.

Everett, Daniel. 2008. *Don't Sleep, There are Snakes: Life and Language in the Amazonian Jungle*. London: Profile.

2009. 'Pirahã Culture and Grammar: A Response to Some Criticism'. *Language* 85(2): 405–42.

Fine, C. 2010. *Delusions of Gender*. New York: W. W. Norton.

Fischhoff, B. 1982. 'Debiasing'. In D. Kahneman, A. Tversky, and P. Slovic (eds.), *Judgment under Uncertainty: Heuristics and Biases*. Cambridge University Press: 422–44.

Fitch, W. T. 2005. 'The Evolution of Language: A Comparative Review'. *Biology and Philosophy* 20: 193–230.

Fitz, H. 2009. 'Neural Syntax'. PhD dissertation. University of Amsterdam. http:// www.illc.uva.nl/Publications/reportlist.php?Series=DS.

Ford, K. M., P. J. Hayes, and C. Glymour. 1998. 'Ramón Lull and the Infidels'. *AI magazine* 19(136).

Frankish, K. 2010. 'Dual-process and Dual-system Theories of Reasoning'. *Philosophy Compass* 5(10): 914–26.

Frege, G. 1879. '*Begriffsschrift*'. In J. van Heijenoort (ed.), *From Frege to Gödel: A Source Book in Mathematical Logic, 1879–1931*. Cambridge, MA: Harvard University Press: 1–82.

 1885. 'Über formale Theorien der Arithmetik'. In E.-H. W. Kluge (ed.), *Frege, Gottlob: On the Foundations of Geometry and Formal Theories of Arithmetic*. New Haven, CT: Yale University Press: 141–53.

Frigg, R., and S. Hartmann. 2006. 'Models in Science'. Ed. E. Zalta. *Stanford Encyclopedia of Philosophy*. http://plato.stanford.edu/entries/models-science/.

Gamut, L. T. F. 1991. *Logic, Language, and Meaning.* Vol. 1. *Introduction to Logic*. University of Chicago Press.

Gibson, K. 2010. 'Talking about Birds, Bees and Primates Too: Implications for Language Evolution'. In A. D. M. Smith (ed.), *The Evolution of Language: Proceedings of the 8th International Conference, Utrecht*. London: World Scientific: 153–9.

Gigerenzer, G. 2007. *Gut Feelings: The Intelligence of the Unconscious*. London: Penguin.

Gigerenzer, G., and D. Goldstein. 1996. 'Reasoning the Fast and Frugal Way: Models of Bounded Rationality'. *Psychological Review* 103: 650–69.

Glymour, C. 1997. *Thinking Things Through*. Cambridge, MA: MIT Press.

Gödel, K. 1995. *Collected Works.* Vol. 3. *Unpublished Essays and Lectures*. Oxford University Press.

Goel, V. 2007. 'The Anatomy of Deductive Reasoning'. *Trends in Cognitive Sciences* 11(10): 435–41.

Goel, V., and R. J. Dolan. 2003. 'Differential Involvement of Left Prefrontal Cortex in Inductive and Deductive Reasoning'. *Cognition* 93: 109–21.

Gopnik, A., and A. Meltzoff. 1997. *Words, Thoughts and Theories*. Cambridge, MA: MIT Press.

Gould, S. J., and E. Vrba. 1982. 'Exaptation: A Missing Term in the Science of Form'. *Paleobiology* 8(1): 4–15.

Grattan-Guiness, I. 2000. *The Search for Mathematical Roots, 1870–1940: Logics, Set Theories, and the Foundations of Mathematics from Cantor through Russell to Gödel*. Princeton University Press.

Griggs, R. A., and J. R. Cox. 1982. 'The Elusive Thematic-Materials Effect in Wason's Selection Task'. *British Journal of Psychology* 73: 407–20.

Harris, R. 1986. *The Origin of Writing*. London: Duckworth.

 1995. *Rethinking Writing*. Harmondsworth: Penguin.

 2009. *Rationality and the Literate Mind*. London: Routledge.

Hauser, M., N. Chomsky, and W. T. Fitch. 2002. 'The Language Faculty: What is it, Who has it, and How did it Evolve?' *Science* 298: 1569–79.

Hécaen, H., and H. Kremin. 1976. 'Neurolinguistic Research on Reading Disorders Resulting from Left Hemisphere Lesions: Aphasic and "Pure" Alexia'. In H. Whitaker and H. A. Whitaker (eds.), *Studies in Neurolinguistics*. New York: Academic Press.

Heeffer, A. 2007. 'Humanist Repudiation of Eastern Influences in Early Modern Mathematics'. http://logica.ugent.be/albrecht/thesis/Repudiation2007.pdf.

2010. 'From the Second Unknown to the Symbolic Equation'. In A. Heeffer and M. Van Dyck (eds.), *Philosophical Aspects of Symbolic Reasoning in Early Modern Mathematics*. London: College Publications: 57–102.

van Heijningen, C. A., J. de Visser, W. Zuidema, and C. ten Cate. 2009. 'Simple Rules Can Explain Discrimination of Putative Recursive Syntactic Structures by a Songbird Species'. *Proceedings of the National Academy of Sciences* 106(48): 20538–43.

Heil, M., B. Rolke, and A. Pecchinenda. 2004. 'Automatic Semantic Activation is No Myth'. *Psychological Science* 15: 852–7.

Henrich, J., S. J. Heine, and A. Norenzayan. 2010. 'The Weirdest People in the World'. *Behavioral and Brain Sciences* 33: 61–83.

Hilbert, D. 1899. 'Grundlagen der Geometrie'. In *Festschrift zur Feier der Enthüllung des Gauss-Weber-Denkmals in Göttingen*. 1st edn. Leipzig: Teubner: 1–92.

Hintikka, J. 1989. 'Is There Completeness in Mathematics after Gödel?' *Philosophical Topics* 17(2): 69–90.

Hobbes, T. 1839. *The English Works Now First Collected and Edited by Sir William Molesworth, Bart.* London: John Bohn.

Hodges, W. 1993. 'The Logical Content of Theories of Deduction'. *Behavioural and Brain Sciences* 16(2): 353–4.

2009. 'Traditional Logic, Modern Logic and Natural Language'. *Journal of Philosophical Logic* 38: 589–606.

Houdé, O., and N. Tzourio-Mazoyer. 2003. 'Neural Foundations of Logical and Mathematical Cognition'. *Nature Reviews Neuroscience* 4: 507–14.

Houdé, O., L. Zago, C. Crivello, S. Moutier, A. Pineau, B. Mazoyer, and N. Tzourio-Mazoyer. 2001. 'Access to Deductive Logic Depends on a Right Ventromedial Prefrontal Area Devoted to Emotion and Feeling: Evidence from a Training Paradigm'. *Neuroimage* 14: 1486–92.

Houdé, O., L. Zago, E. Mellet, S. Moutier, A. Pineau, B. Mazoyer, and N. Tzourio-Mazoyer. 2000. 'Shifting from the Perceptual Brain to the Logical Brain: The Neural Impact of Cognitive Inhibition Training'. *Journal of Cognitive Neuroscience* 12: 721–8.

Høyrup, J. 2006. 'Pre-modern Algebra: A Concise Survey of That Which Was Shaped into the Technique and Discipline We Know'. In B. Henning, V. Hendricks, F. Voetmann, and K. Christiansen (eds.), *Essays in Honour of Stig Andur Pedersen*. London: College Publications: 1–15.

2007. 'Generosity: No Doubt, but at Times Excessive and Delusive'. *Journal of Indian Philosophy* 35: 469–85.

2010. 'Hesitating Progress: The Slow Development toward Algebraic Symbolization in Abbacus- and Related Manuscripts, c.1300 to c.1550'. In A. Heeffer and

M. Van Dyck (eds), *Philosophical Aspects of Symbolic Reasoning in Early Modern Mathematics*. London: College Publications: 3–56.

Hubien, H. (ed.) 1976. *Iohannis Buridani Tractatus de consequentiis*. Philosophes Médiévaux XVI. Louvain: Publications universitaires.

Immerman, N. 2004. 'Computability and Complexity'. Ed. E. Zalta. *Stanford Encyclopedia of Philosophy*. http://plato.stanford.edu/entries/computability/.

Inhelder, B., and J. Piaget. 1958. *The Growth of Logical Thinking from Childhood to Adolescence*. New York: Basic Books.

Jackson, Frank. 1991. *Conditionals* (Oxford Readings in Philosophy). Oxford University Press.

Jobard, G., F. Crivello, and N. Tzourio-Mazoyer. 2003. 'Evaluation of the Dual Route Theory of Reading: A Metanalysis of 35 Neuroimaging Studies'. *NeuroImage* 20: 693–712.

Johnson-Laird, P. N. 2008. 'Mental Models and Deductive Reasoning'. In J. Adler and L. Rips (eds.), *Reasoning: Studies of Human Inference and its Foundations*. Cambridge University Press: 206–22.

Johnson-Laird, P. N., M. S. Legrenzi, and P. Legrenzi. 1972. 'Reasoning and a Sense of Reality'. *British Journal of Psychology* 63: 395–400.

Keren, G., and Y. Schul. 2009. 'Two is Not Always Better than One: A Critical Evaluation of Two-system Theories'. *Perspectives on Psychological Science* 4(6): 533–50.

Kirby, S. 2012. 'Language is an Adaptive System: The Role of Cultural Evolution in the Origins of Structure'. In *Oxford Handbook of Language Evolution*. Oxford University Press: 589–604.

Kirby, S., K. Smith, and H. Brighton. 2004. 'From UG to Universals: Linguistic Adaptation through Iterated Learning'. *Studies in Language* 28(3): 1–17.

Kirsh, D. 2010. 'Thinking with External Representations'. *AI and Society* 25: 441–54.

Klayman, J. 1995. 'Varieties of Confirmation Bias'. *The Psychology of Learning and Motivation* 32: 385–419.

Kleene, S. 1951. *Introduction to Metamathematics*. Amsterdam: North Holland.

Krämer, S. 1991. *Berechenbare Vernunft: Kalkül und Rationalismus im 17. Jahrhundert*. Berlin: Walter de Gruyter.

2003. 'Writing, Notational Iconicity, Calculus: On Writing as a Cultural Technique'. *Modern Languages Notes (German Issue)* 118(3): 518–37.

Laidler, K. 1998. *To Light Such a Candle*. Oxford University Press.

Lakoff, G., and R. Nuñez. 2000. *Where Mathematics Comes From*. New York: Basic Books.

Lam, L. 1996. 'The Development of Hindu–Arabic and Traditional Chinese Arithmetic'. *Chinese Science* 13: 35–54.

Landy, D., and R. L. Goldstone. 2007a. 'Formal Notations are Diagrams: Evidence from a Production Task'. *Memory and Cognition* 35(8): 2033–40.

2007b. 'How Abstract is Symbolic Thought?' *Journal of Experimental Psychology: Learning, Memory, and Cognition* 33(4): 720–33.

2009. 'Pushing Symbols'. In *Proceedings of the 31st Annual Conference of the Cognitive Science Society, Amsterdam*. Austin, TX: Cognitive Science Society.

Larrick, R. P. 2004. 'Debiasing'. In Derek J. Koehler and Nigel Harvey (eds.), *Blackwell Handbook of Judgment and Decision Making*. Oxford University Press: 316–37.

Legg, C. 2011. 'The Hardness of the Iconic Must: Can Peirce's Existential Graphs Assist Modal Epistemology?' doi:10.1093/philmat/nkr005.

Leibniz, G. W. 1966. *Logical Papers: A Selection*. Oxford: Clarendon Press.

  1982. *Vorausedition zur Reihe VI: Philosophische Schriften*. Münster: Leibnizforschungsstelle.

  1989. *Philosophical Essays*. Indianapolis, IN: Hackett Publishing Company.

  2000. *Die Grundlagen des logischen Kalküls*. Hamburg: Felix Meiner.

Lenzen, W. 2004. 'Leibniz's Logic'. In D. Gabbay and J. Woods (eds.), *Handbook of the History of Logic*, Vol. 3. *The Rise of Modern Logic: From Leibniz to Frege*. Amsterdam: Elsevier: 1–83.

Levesque, H. J., and R. J. Brachman. 1987. 'Expressiveness and Tractability in Knowledge Representation and Reasoning'. *Computational Intelligence Journal* 3: 78–93.

Lewis, David. 1976. 'Probabilities of Conditionals and Conditional Probabilities'. *Philosophical Review* 85: 297–315.

Linell, P. 2005. *The Written Language Bias in Linguistics: Its Nature, Origins and Transformations*. London: Routledge.

Lloyd, G. E. R. 1996. 'Science in Antiquity: The Greek and Chinese Cases and their Relevance to the Problem of Culture and Cognition'. In D. Olson and N. Torrance (eds.), *Modes of Thought: Explorations in Culture and Cognition*. Cambridge University Press: 15–33.

Loeffler, R. 2009. 'Neo-Pragmatist (Practice-Based) Theories of Meaning'. *Philosophy Compass* 4: 197–218.

Lohr, C. 1974. Review of J. N. Hillgarth, *Ramon Lull and Lullism in Fourteenth-century France*. *Speculum* 49(1): 121–4.

  2010. 'Ramon Lull (1232–1316): The Activity of God and the Hominization of the World'. In P. R. Blum (ed.), *Philosophers of the Renaissance*. Washington, DC: Catholic University of America Press.

Luria, A. R. 1976. *Cognitive Development: Its Social and Cultural Foundations*. Cambridge, MA: Harvard University Press.

Maat, J. 2004. *Philosophical Languages in the Seventeenth Century: Dalgarno, Wilkins, Leibniz*. New Synthese Historical Library, 54. Dordrecht: Kluwer.

Macbeth, D. 2004. 'Viète, Descartes, and the Emergence of Modern Mathematics'. *Graduate Faculty Philosophy Journal* 25(2): 87–117.

  2011. 'Seeing How it Goes: Paper-and-Pencil Reasoning in Mathematical Practice', *Philosophia Mathematica* 20(1): 58–85.

  forthcoming. 'Writing Reason'. *Logique et Analyse*.

MacFarlane, J. 2000. 'What Does it Mean to Say that Logic is Formal?' PhD dissertation. University of Pittsburgh. http://fitelson.org/confirmation/macfarlane_ch_7.pdf.

Macnamara, J. 1986. *A Border Dispute: The Place of Logic in Psychology*. Cambridge, MA: MIT Press.

Manktelow, K. I., and J. Evans. 1979. 'Facilitation of Reasoning by Realism: Effect or Not-effect'. *British Journal of Psychology* 70: 477–88.

Marion, M., and B. Castelnerac. 2009. 'Arguing for Inconsistency: Dialectical Games in the Academy'. In G. Primiero and S. Rahman (eds.), *Acts of Knowledge: History, Philosophy and Logic*. London: College Publications: 37–76.

Marr, D. 1982. *Vision: A Computational Investigation into the Human Representation and Processing of Visual Information*. San Francisco, CA: W. H. Freeman.

Menary, R. 2007a. *Cognitive Integration: Mind and Cognition Unbounded*. Basingstoke: Palgrave Macmillan.

2007b. 'Writing as Thinking'. *Language Sciences* 29: 621–32.

2010a. 'Dimensions of Mind'. *Phenomenology and the Cognitive Sciences* 9(4): 561–78.

(ed.) 2010b. *The Extended Mind*. Cambridge, MA: MIT Press.

Mercier, H., and D. Sperber. 2011. 'Why Do Humans Reason? Arguments for an Argumentative Theory'. *Behavioral and Brain Sciences* 34(2): 57–74.

Meyer, D. E., and R. W. Schvaneveldt. 1971. 'Facilitation in Recognizing Pairs of Words: Evidence of a Dependence between Retrieval Operations'. *Journal of Experimental Psychology*, 90, 227–34.

Milne, P. 2003. 'The Simplest Lewis-style Triviality Proof Yet?' *Analysis* 63: 300–3.

Morgan, J. J., and J. T. Morton. 1944. 'The Distortion of Syllogistic Reasoning Produced by Personal Convictions'. *Journal of Social Psychology* 20: 39–59.

Morris, A. K., and V. Sloutsky. 1998. 'Understanding of Logical Necessity: Developmental Antecedents and Cognitive Consequences'. *Child Development* 69(3): 721–41.

Moshman, G., and M. Geil. 1998. 'Collaborative Reasoning: Evidence for Collective Rationality'. *Thinking and Reasoning* 4: 231–48.

Moutier, S., N. Angeard, and O. Houdé. 2002. 'Deductive Reasoning and Matching-bias Inhibition Training: Evidence from a Debiasing Paradigm'. *Thinking and Reasoning* 8(3): 205–24.

Moutier, S., S. Plagne, A.-M. Melot, and O. Houdé. 2006. 'Syllogistic Reasoning and Belief-bias Inhibition in School Children'. *Developmental Science* 9: 166–72.

Mugnai, M. 2010. 'Logic and Mathematics in the Seventeenth Century'. *History and Philosophy of Logic* 31: 297–314.

Neely, J. H. 1991. 'Semantic Priming Effects in Visual Word Recognition: A Selective Review of Current Findings and Theories'. In Derek Besner and Glyn W. Humphreys (eds.), *Basic Processes in Reading: Visual Word Recognition*. Hillsdale, NJ: L. Erlbaum Associates: 264–336.

Neely, J. H., and T. A. Kahan. 2001. 'Is Semantic Activation Automatic? A Critical Re-evaluation'. In A. M. Surprenant, I. Neath, H. L. Roediger, and J. S. Nairne (eds.), *The Nature of Remembering: Essays in Honor of Robert G. Crowder*. Washington, DC: American Psychological Association.

Nersessian, N. 1984. *Faraday to Einstein: Constructing Meaning in Scientific Theories*. Dordrecht: Martinus Nijhoff.

Netz, R. 1999. *The Shaping of Deduction in Greek Mathematics: A Study in Cognitive History*. Cambridge University Press.

2012. 'Reasoning and Symbolism in Diophantus: Preliminary Observations'. In K. Chemla (ed.), *The History of Mathematical Proof in Ancient Traditions*. Cambridge University Press.

Nevins, A., C. Rodrigues, and D. Pesetsky. 2009. 'Evidence and Argumentation: A Reply to Everett'. *Language* 85(3): 671–81.

De Neys, W., and E. Van Gelder. 2009. 'Logic and Belief across the Lifespan: The Rise and Fall of Belief Inhibition during Syllogistic Reasoning'. *Developmental Science* 12: 123–30.

Nickerson, R. 1998. 'Confirmation Bias: A Ubiquitous Phenomenon in Many Guises'. *Review of General Psychology* 2(2): 175–220.

Niyogi, P., and R. C. Berwick. 1997. 'A Dynamical Systems Model for Language Change'. *Complex Systems* 11: 161–204.

Oakhill, J. V., and P. N. Johnson-Laird. 1985. 'The Effect of Belief on the Production of Syllogistic Conclusions'. *Quarterly Journal of Experimental Psychology* 37A: 553–69.

Oaksford, M. R., and N. C. Chater. 1994. 'A Rational Analysis of the Selection Task as Optimal Data Selection'. *Psychological Review* 101: 608–31.

2002. 'Commonsense Reasoning, Logic and Human Rationality'. In R. Elio (ed.), *Commonsense Reasoning and Rationality*. Oxford University Press: 174–214.

Oaksford, M. R., U. Hahn, and N. C. Chater. 2008. 'Human Reasoning and Argumentation: The Probabilistic Approach'. In J. Adler and L. Rips (eds.), *Reasoning: Studies of Human Inference and its Foundation*. Cambridge University Press: 383–413.

Oller, D. K., and U. Griebel. 2006. 'How the Language Capacity was Naturally Selected: Altriciality and Long Immaturity'. *Behavioral and Brain Sciences* 29(3): 293–4.

Olson, D. 1994. *The World on Paper: The Conceptual and Cognitive Implications of Writing and Reading*. Cambridge University Press.

Over, D. (ed.) 2003. *Evolution and the Psychology of Thinking: The Debate*. Hove: The Psychology Press.

Over, D., and J. Evans. 2003. 'The Probability of Conditionals: The Psychological Evidence'. *Mind and Language* 18(4): 340–58.

Pagin, P., and D. Westerståhl. 2010. 'Compositionality I: Definitions and Variants'. *Philosophy Compass* 5: 250–64.

Paoli, F. 2003. 'Quine and Slater on Paraconsistency and Deviance'. *Journal of Philosophical Logic* 32: 531–48.

Papineau, D. 2007. 'Naturalism'. Ed. E. Zalta. *Stanford Encyclopedia of Philosophy*. http://plato.stanford.edu/entries/naturalism/.

Peckhaus, V. 1997. *Logik, Mathesis universalis und allgemeine Wissenschaft: Leibniz und die Wiederentdeckung der formalen Logik im 19. Jahrhundert*. Berlin: Akademie Verlag.

2004. 'Calculus Ratiocinator Versus Characteristica Universalis? The Two Traditions in Logic, Revisited'. *History and Philosophy of Logic* 25(1): 3–14.

2009. 'Leibniz's Influence on Nineteenth Century Logic'. Ed. E. Zalta. *The Stanford Encyclopedia of Philosophy*. http://plato.stanford.edu/entries/leibniz-logic-influence/.

Peirce, C. S. 1931. *Collected Writings*. Cambridge, MA: Harvard University Press.

Pelletier, F. J. 1991. 'The Philosophy of Automated Theorem Proving'. In J. Mylopoulos and R. Reiter (eds.), *Proceedings of the 12th International Joint Conference on Artificial Intelligence, Darling Harbour, Sydney, Australia, 24–30 August 1991*. 2 Vols. San Mateo, CA: Morgan Kaufmann. Vol. 2, pp. 1039–45.

1998. 'Natural Deduction Theorem Proving in THINKER'. *Studia Logica* 60: 3–43.

Pinker, S., and P. Bloom. 1990. 'Natural Language and Natural Selection'. *Behavioral and Brain Science* 13: 707–84.

Prado, J., and I. A. Noveck. 2007. 'Overcoming a Perceptual Bias: A Parametric fMRI Study'. *Journal of Cognitive Neuroscience* 19(4): 642–57.

Priest, G. 2006. *Doubt Truth to be a Liar*. Oxford University Press.

Prowse Turner, J. A., and V. A. Thompson. 2009. 'The Role of Training, Alternative Models, and Logical Necessity in Determining Confidence in Syllogistic Reasoning'. *Thinking and Reasoning* 15(1): 69–100.

Quine, W. V. O. 1953. 'Reference and Modality'. In W. V. O. Quine, *From a Logical Point of View*, Cambridge, MA: Harvard University Press: 139–59.

Ramsey, F. P. 1929. 'General Propositions and Causality'. In D. H. Mellor (eds.), *F. P. Ramsey: Philosophical Papers*. Cambridge University Press.

Rendell, L., and H. Whitehead. 2001. 'Culture in Whales and Dolphins'. *Behavioral and Brain Science* 24: 309–82.

Ridley, M. 2003. *Nature via Nurture: Genes, Experience, and What Makes us Human*. New York: HarperCollins.

Rips, L. 1994. *The Psychology of Proof*. Cambridge, MA: MIT Press.

2002. 'Reasoning Imperialism'. In R. Elio (eds.), *Common Sense, Reasoning, and Rationality*. Oxford University Press: 215–35.

2008. 'Logical Approaches to Human Deductive Reasoning'. In J. Adler and L. Rips (eds.), *Reasoning: Studies of Human Inference and its Foundations*. Cambridge University Press: 187–205.

Sá, W., R. F. West, and K. E. Stanovich. 1999. 'The Domain Specificity and Generality of Belief Bias: Searching for a Generalizable Critical Thinking Skill'. *Journal of Educational Psychology* 91: 497–510.

Sacharov, A. 1999. 'Formal Language'. *MathWorld*. http://mathworld.wolfram.com/FormalLanguage.html.

Schlimm, D. 2011. 'On the Creative Role of Axiomatics. The Discovery of Lattices by Schröder, Dedekind, Birkhoff, and Others'. *Synthese* 183(1): 47–68.

Schlimm, D., and H. Neth. 2008. 'Modeling Ancient and Modern Arithmetic Practices: Addition and Multiplication with Arabic and Roman Numerals'. In V. Sloutsky, B. Love, and K. McRae (eds.), *30th Annual Conference of the Cognitive Science Society*. Austin, TX: Cognitive Science Society: 2097–102.

Schmandt-Besserat, D. 1996. *How Writing Came About*. Austin, TX: University of Texas Press.

Scribner, S. 1997. *Mind and Social Practice*. Cambridge University Press.

References

Serfati, M. 2005. *La Révolution symbolique: la constitution de l'écriture symbolique mathématique*. Paris: Éditions Pétra.

Shapiro, S. 1998. 'Logical Consequence: Models and Modality'. In M. Schirn (ed.), *Philosophy of Mathematics Today*. Oxford: Clarendon Press: 131–56.

2006. 'Logical Consequence, Proof Theory, and Model Theory'. In S. Shapiro (ed.), *Oxford Handbook of Philosophy of Mathematics and Logic*. Oxford University Press: 651–70.

Shin, S.-J., and O. Lemon. 2001. 'Diagrams'. Ed. E. Zalta. *Stanford Encyclopedia of Philosophy*. http://plato.stanford.edu/entries/diagrams/.

Shoham, Y. 1987. 'A Semantical Approach to Non-monotonic Logics', Proceedings of the Tenth International Joint Conference on Artificial Intelligence, Milan, Italy: 388–92. Reprinted in M. L. Ginsberg (ed.), *Readings in Nonmonotonic Reasoning* (Los Altos, CA: Morgan Kaufmann, 1987).

Sieg, W. 1994. 'Mechanical Procedures and Mathematical Experiences'. In A. George (ed.), *Mathematics and Mind*. Oxford University Press: 71–117.

2006. 'Gödel on Computability'. *Philosophia Mathematica* 14: 189–207.

2008. 'On Computability'. In A. Irvine (ed.), *Handbook of the Philosophy of Science: Philosophy of Mathematics*. Amsterdam: Elsevier: 525–621.

Smith, P. 2009. *An Introduction to Gödel's Theorems*. Cambridge University Press.

2010. 'Squeezing Arguments'. *Analysis* 71(1): 22–30.

Smith, R. (ed.) 1989. *Aristotle: Prior Analytics*. Indianapolis, IN: Hackett Publishing Company.

Staal, F. 2006. 'Artificial Languages across Sciences and Civilizations'. *Journal of Indian Philosophy* 34: 89–141.

2007a. 'Artificial Languages between Innate Faculties'. *Journal of Indian Philosophy* 35: 577–96.

2007b. 'Preface: The Generosity of Formal Languages'. *Journal of Indian Philosophy* 35: 405–12.

Stalnaker, R. 1970. 'Probability and Conditionals'. *Philosophy of Science* 37: 64–80.

Stanovich, K. E. 1999. *Who Is Rational? Studies of Individual Differences in Reasoning*. Mahwah, NJ: Lawrence Erlbaum Associates.

2003. 'The Fundamental Computational Biases of Human Cognition: Heuristics that (Sometimes) Impair Decision Making and Problem Solving'. In J. E. Davidson and R. J. Sternberg (eds.), *The Psychology of Problem Solving*. Cambridge University Press: 291–342.

2004. *The Robot's Rebellion: Finding Meaning in the Age of Darwin*. University of Chicago Press.

2008. 'Individual Differences in Reasoning and the Algorithmic/Intentional Level'. In J. Adler and L. Rips (eds.), *Reasoning: Studies of Human Inference and its Foundations*. Cambridge University Press: 414–36.

2009. 'Distinguishing the Reflective, Algorithmic, and Autonomous Minds: Is it Time for a Tri-process Theory?' In J. Evans and K. Frankish (eds.), *In Two Minds: Dual Processes and Beyond*. Oxford University Press: 55–88.

Steedman, M. 2002. 'Plans, Affordances, and Combinatory Grammar'. *Linguistics and Philosophy* 25(5–6): 723–53.

Stein, H. 1970. 'On the Notion of Field in Newton, Maxwell, and Beyond'. In R. H. Stuewer (ed.), *Historical and Philosophical Perspectives of Science*. Minnesota Studies in the Philosophy of Science, Vol. 5. Minneapolis, MN: University of Minnesota Press: 264–87.

1981. '"Subtler Forms of Matter" in the Period Following Maxwell'. In G. N. Cantor and M. J. S. Hodge (eds.), *Conceptions of Ether: Studies in the History of Ether Theories, 1740–1900*. Cambridge University Press: 309–40.

Stenning, K. 2002. *Seeing Reason: Image and Language in Learning to Think*. Oxford University Press.

Stenning, K., and M. van Lambalgen. 2004. 'A Little Logic Goes a Long Way: Basing Experiment on Semantic Theory in the Cognitive Science of Conditional Reasoning'. *Cognitive Science* 28: 481–529.

2008. *Human Reasoning and Cognitive Science*. Cambridge, MA: MIT Press.

Stokhof, M. 2007. 'Hand or Hammer? On Formal and Natural Languages in Semantics'. *Journal of Indian Philosophy* 35: 597–626.

Stolz, J. A., and D. Besner. 1999. 'On the Myth of Automatic Semantic Activation in Reading'. *Current Directions in Psychological Science* 8: 61–4.

Stroop, J. R. 1935. 'Studies of Interference in Serial Verbal Reactions'. *Journal of Experimental Psychology* 18: 643–61.

Sutton, J. 2010. 'Exograms and Interdisciplinarity: History, the Extended Mind, and the Civilizing Process'. In R. Menary (ed.), *The Extended Mind*. Cambridge, MA: MIT Press: 189–226.

Tarski, A. 1944. 'The Semantic Conception of Truth and the Foundations of Semantics'. *Philosophy and Phenomenological Research* 4: 341–76.

1959. *Introduction to Logic and to the Methodology of Deductive Sciences*. New York: Oxford University Press.

Tomasello, M. 2003. *Constructing a Language: A Usage-Based Theory of Language Acquisition*. Cambridge, MA: Harvard University Press.

2008. 'How are Humans Unique?' *New York Times Magazine*, 25 May 2008.

Torrens, D., V. A. Thompson, and K. M. Cramer. 1999. 'Individual Differences and the Belief Bias Effect: Mental Models, Logical Necessity, and Abstract Reasoning'. *Thinking and Reasoning* 5: 1–28.

Turing, A. M. 1936. 'On Computable Numbers, with an Application to the *Entscheidungsproblem*'. *Proceedings of the London Mathematical Society* 2(42): 230–65.

Veblen, O. 1906. 'The Foundations of Geometry: A Historical Sketch and a Simple Example'. *Popular Science Monthly* 68: 21–8.

Wang, H. 1955. 'On Formalization'. *Mind* 64: 226–38.

1990. *Computation, Logic, Philosophy: A Collection of Essays*. Dordrecht: Kluwer.

Wason, P. C. 1966. 'Reasoning'. In B. Foss (ed.), *New Horizons in Psychology*. Harmondsworth: Penguin: 135–51.

Wason, P. C., and D. W. Green. 1984. 'Reasoning and Mental Representation'. *Quarterly Journal of Experimental Psychology* 36A: 597–610.

Wason, P. C., and D. Shapiro. 1971. 'Natural and Contrived Experience in a Reasoning Problem'. *Quarterly Journal of Experimental Psychology* 23: 63–71.

Weir, A. 2011. 'Formalism in the Philosophy of Mathematics'. Ed. E. Zalta. *Stanford Encyclopedia of Philosophy*. http://plato.stanford.edu/entries/formalism-mathematics/.

Whitehead, A. 1911. *An Introduction to Mathematics*. Oxford University Press.

Wilkins, M. C. 1928. 'The Effect of Changed Material on Ability to do Formal Syllogistic Reasoning'. *Archives of Psychology* 16: 1–83.

Wilson, M. 2002. 'Six Views of Embodied Cognition'. *Psychonomic Bulletin and Review* 9: 625–36.

Wittgenstein, L. 1953. *Philosophical Investigations*. Oxford: Blackwell.

    1963. *Tractatus Logico-Philosophicus*. London: Routledge and Kegan Paul.

Von Wright, G. 1951. *An Essay in Modal Logic*. Amsterdam: North-Holland.

Zach, R. 2003. 'Hilbert's Program'. Ed. E. Zalta. *Stanford Encyclopedia of Philosophy*. http://plato.stanford.edu/entries/hilbert-program/.

    2006. 'Kurt Gödel and Computability Theory'. In A. Beckmann, U. Berger, B. Löwe, and J. V. Tucker (eds.), *Logical Approaches to Computational Barriers*. Second Conference on Computability in Europe, CiE 2006, Swansea, UK, June/July 2006. Proceedings. Berlin: Springer Verlag: 575–83.

Index

algebra, 73, 80, 82, 227
  abbaco schools, 67, 75, 86
  al-Khwārizmī, 21, 72, 73, 74, 75, 76, 78
  algebraic approach to logic, 79, 80, 82, 83
  algebraic notations, 73, 74, 83, 84, 98, 109
  Diophantus of Alexandria, 73, 74, 76
algorithm, 21, 25, 27, 63, 72, 230
Aristotle, 51, 67, 152, 251
  conception of nature, 46
  Prior Analytics, 69, 71, 77
  syllogistic, 81, 83
arithmetic, 14, 19, 21, 59, 75, 79, 84, 86, 101, 104, 105, 106, 107, 176, 206, 223, 224
axiomatic systems, 48, 59, 85, 86, 87, 205, 215, 216, 224

Boole, George, 82, 83, 84, 94, 99

calculation, 23, 62, 73, 75, 77, 78, 79, 80, 81, 87, 94, 95, 108
calculus, 55, 200, 216, 227, 231
  probability, 219
Carnap, Rudolf, 15, 29, 64, 90, 96, 101, 171, 203, 204, 232
Church thesis, 26
Clark, Andy, 6, 56, 162, 180, 181, 182, 183, 191, 192, 196
closed world assumption, 142
closed world reasoning, 127, 142, 150
cognitive integration, 6, 177, 184, 193, 201
cognitive manipulations, 183, 184, 193, 194, 202
cognitive monitoring, 240, 241, 242, 248
complementarity principle, 192
computable
  concept of, 21, 22, 25, 27
  not requiring insight or ingenuity, 21, 23, 25, 26
conditionals, 125, 128, 216, 238
  defeasible, 124, 125, 127, 129, 142, 157
  deontic, 121, 122, 123, 124

probabilistic, 119, 124, 217, 218
semantics of, 217, 218

d'Alembert, Jean, 7, 227
de-semantification, 54, 59, 86, 137, 198, 199, 201, 202, 203, 210, 215, 216, 219, 222, 228, 242, 245
  cognitive implications, 207, 210
  epistemic freedom, 200
  metaphysical freedom, 200
debiasing, 64, 137, 152, 153, 159, 173, 185, 193, 197, 204, 211, 221, 222, 227, 228, 231, 232, 235, 240, 241, 242, 245
Dedekind, Richard, 19, 85, 104, 107, 226
deductive method, 16, 68, 149, 153, 154, 155, 156, 159, 160, 222, 231
Dehaene, Stanislas, 33, 43, 149, 158, 161, 163, 165, 166, 168, 173, 174, 176, 177
  neuronal recycling, 176, 177
Descartes, René, 22, 23, 75, 76
dialogical conception of deduction, 2, 68, 154, 155, 156, 160
doxastic conservativeness, 134, 136, 146, 149, 150, 151, 153, 154, 160, 211, 212, 215, 221, 222, 228, 230, 233, 247, 255
dual-process theories of cognition, 157, 158, 159, 236, 242, 243, 243, 245, 246, 247, 248
  type 1 v. type 2, 243, 244

Entscheidungsproblem, 17
Evans, Jonathan, 5, 119, 120, 121, 131, 132, 133, 136, 141, 147, 148, 157, 159, 210, 211, 236, 237, 238, 242, 244, 245
exograms, 192, 193, 195
extended cognition, 27, 60, 67, 161, 178, 180, 181, 182, 186, 190, 191, 200
  pathologies of, 195
extended mind hypothesis, 162, 180, 182, 191
externalization of reasoning, 23, 27, 60, 162, 194, 196, 197, 216, 228, 245, 248

273

formal
  as computable, 12, 16, 17, 19, 20, 27, 60,
    194, 244
  as de-semantification, 12, 13, 14, 15, 16, 59, 60,
    86, 96, 198, 203
  as mechanical rule following, 17, 18, 19, 230
  as pertaining to forms, 12
  as pertaining to rules, 12
  as schematic, 72, 88, 114, 116
formal languages
  as a technology, 3, 61, 66, 193, 255
  as cognitive artefacts, 2, 27, 65, 91, 96, 100, 172,
    173, 181, 183, 197, 255
  as constitutive of reasoning processes, 161, 181,
    183, 185, 189, 190, 197
  as mathematical objects, 1, 29, 52, 58, 203
  diagrammatic, 53, 83, 93, 186, 187
  limitations, 66, 97, 100, 103, 104, 105, 106, 108
  rationales for, 63, 64, 66, 89, 90, 91, 92, 93, 94,
    95, 96, 97, 99, 104, 167, 203
formal system, 18, 19, 58, 101, 104, 105, 106
formalization, 223, 231, 235
  adequacy of, 226, 227, 229, 230, 234, 235
  as conceptual translation, 224
  different stages, 226, 227, 230
  epistemic access to target, 223, 224
  target phenomenon of, 97, 99, 101, 102, 104, 105,
    206, 223, 224, 225, 226, 227, 228, 232, 233, 235
Frege, Gottlob, 5, 14, 16, 19, 20, 46, 63, 64, 84, 85,
    87, 89, 90, 92, 93, 96, 99, 100, 107, 153, 189,
    198, 202, 205, 206, 228, 229, 251
  Begriffsschrift, 47, 84, 85, 89, 90, 96, 198, 203

Gigerenzer, Gerd, 147, 150, 151, 245
Gödel, Kurt, 17, 18, 21, 87, 105

Harris, Roy, 41, 42
Hilbert, David, 13, 14, 15, 16, 17, 20, 86, 87, 101,
    107, 198, 203, 204, 205, 206, 219, 228
Hindu–Arabic numerals, 27, 62, 63
Hobbes, Thomas, 79, 80
Hypothetical Thinking Theory (HTT), 141

inhibition training, 7, 152, 158, 222, 236, 239, 240,
    242, 243, 245

Kant, Immanuel, 15, 114, 147, 253
Krämer, Sybille, 6, 13, 25, 41, 44, 53, 54, 55, 56,
    59, 62, 63, 79, 95, 162, 190, 198, 199, 200,
    201, 202

Landy, David, 6, 67, 93, 162, 175, 184, 188
languages
  artificial v. natural, 45, 46, 47, 48, 49, 50
  as cultural evolution, 36

communicative function, 37, 38, 39, 40,
    55, 56
  evolutionary emergence of, 31, 33, 36, 37, 38, 39,
    40, 49
  main characteristics of human, 30, 57
  regimentation of, 68, 69, 70, 71, 77, 88, 91
  semantics, 32, 33, 34, 54
  speech, 31, 32, 53
  spoken v. written, 35, 40, 41, 43, 44, 45, 47,
    50, 69
  syntax, 34, 35, 36, 37, 54
Leibniz, Gottfried, 5, 20, 23, 64, 68, 76, 79, 80, 81,
    82, 83, 84, 91, 94, 200, 206, 230, 248, 253
Lewis, David, 199, 216, 217, 218, 219, 232
logical consequence, 18, 90, 139, 140, 143
logical constants, 98, 99, 231, 232
logical form, 122, 126, 207
Lull, Ramon, 22, 68, 79

machines as metaphor for reasoning, 18, 22, 23, 24,
    79, 244
Marr, David, 113, 119, 128, 229
mathematical notations, 41, 62, 66, 67, 71, 73, 74,
    75, 76, 77, 80, 81, 84, 86, 87, 172
  history of, 172, 199
Maxwell, James Clerk, 99, 199, 204, 212, 213, 214,
    215, 216, 220, 223, 225
Menary, Richard, viii, 6, 56, 177, 178, 180, 182, 183,
    184, 188, 191, 193, 194, 195, 201, 202
mental models, 118, 119, 126, 138, 139
mental rules, 114, 115, 118, 119, 126, 140
meta-cognition, 240, 241, 242, 245, 247, 248
metalogic, 85, 86, 87, 100, 104, 203,
    231, 232
methodology in philosophy, 7, 152
  conceptual archaeology, 252
  conceptual reflection, 250, 256
  conjunctive methodological pluralism, 255, 256
  empirically informed, 2, 3, 249, 253, 253, 254,
    256, 257
  experimental philosophy, 253
  formal methods, 249, 251
  historically informed, 2, 3, 252, 252, 254, 256
  integrative, 249, 255
  intuitions, 251, 254
  modal v. amodal, 162, 166, 184, 198
  monotonicity v. non-monotonicity, 124, 125, 127,
    128, 140, 142, 143, 148, 157, 160, 222

Netz, Raviel, 3, 66, 68, 71, 74, 154, 155, 249

parity principle, 191, 192
Peirce, Charles Sanders, 83, 93
Piaget, Jean, v, 114, 115, 116, 236, 237
place–value numerical systems, 72, 73, 201

planning, 140, 149, 157, 246
preferred models, 141, 143, 144, 145, 146, 153, 159, 160
priming, 208, 209, 241
    subliminal, 209
Principia Mathematica, 85, 87, 108
probabilistic approaches to reasoning, 119

Ramsey, Frank, 119, 217, 218, 225, 269
rationality, 147, 151, 154, 222, 237, 246
reading
    dual-route model, 162, 164, 165, 166, 167, 170
    letters v. numbers, 162, 168, 169
    neuroscience of, 158, 161, 162, 163, 164, 166, 168, 170, 173, 174, 176, 178
reasoning as defeasible, 125, 126, 127, 128
reasoning biases, 130, 221, 222, 240
    belief bias, 131, 132, 134, 135, 136, 137, 138, 144, 146, 148, 149, 151, 152, 153, 207, 214, 228, 238
    'fundamental computational bias', 64, 129, 131, 137, 141, 151, 196, 246
    matching bias, 117, 210, 211, 236, 238, 239, 240
    'unfamiliar content', 136
re-semantification, vi, 6, 199, 203, 204, 205, 206, 211, 216, 219
rules of formation v. rules of transformation, 19, 58, 59, 60, 81, 84, 87, 90, 174, 175, 194, 203

schematic letters, 68, 69, 70, 71, 77, 88, 118
scientific contexts, 47, 64, 90, 149, 151, 152, 154, 160, 222
semantic activation, 6, 33, 199, 206, 207, 208, 209, 210, 211, 219, 228, 238, 242, 245, 247
sensorimotor processing of notations, 28, 60, 162, 175, 185, 188, 189, 190, 193, 198, 222, 228, 238, 241, 245
squeezing argument, 21, 227, 233, 234, 235
Stanovich, Keith, 5, 64, 113, 129, 130, 131, 136, 137, 141, 144, 150, 151, 196, 221, 236, 246

Stenning, Keith, viii, 6, 39, 40, 49, 54, 98, 113, 115, 118, 119, 123, 124, 126, 127, 128, 129, 139, 140, 141, 142, 149, 151, 155, 156, 162, 184, 186, 187, 195, 246
Stroop effect, 208, 209
suppression task (Byrne), 125, 126
syllogistic, 70, 77, 82, 131, 132, 134, 135, 137, 148, 207
system imprisonment, 100, 101, 102, 103, 109, 196

Tomasello, Michael, 32, 33, 34, 35, 51, 55, 156
Turing, Alan, 17, 18, 21, 26
Turing machine, 23, 26, 27, 97

van Lambalgen, Michiel, vii, 39, 113, 116, 118, 119, 123, 124, 126, 127, 129, 138, 139, 140, 142, 147, 149, 150, 151, 246
Viète, François, 67, 74, 75, 76, 79, 84, 200

Wason selection task, 115, 116, 117, 120, 121, 124, 211, 238, 239
    contentual material, 117, 120, 121
    in groups, 157
    unfamiliar material, 122
Whitehead, Alfred, 28, 85, 96, 185, 186, 241, 244
Wittgenstein, Ludwig, 5, 24, 30, 32, 51, 92, 96, 229
writing, 175
    as cultural product, 149
    formal languages as, 42, 53, 56, 62, 164, 167, 169, 171, 175
    history of, 42, 43, 62
    operative, 55, 56, 73, 162, 169, 178, 184, 200, 201, 202
    phonographic conception, 41, 42, 53, 56, 164, 167
written language bias, 35, 44

Printed in the United States
By Bookmasters